Holt Math
State Test Prep
Workbook

Grade 9

HOLT, RINEHART AND WINSTON

A Harcourt Education Company

Orlando • **Austin** • New York • San Diego • London

ISBN 0-03-077948-0

2 3 4 5 6 7 8 9 862 09 08 07 06

Contents

Grade 9

Name _____ Date _____ Class _____

Diagnostic Assessment
Number Properties and Operations

Solve each problem. Choose the best answer for each question and record your answer on the Student Answer Sheet.

1. The Berry Company has $43,500,000 in total sales for the year. What is this number in scientific notation?

   A 43×10^6

   B 435×10^6

   C 4.35×10^5

   D 4.35×10^7

   E 4.35×10^6

2. Which equation represents the statement "the length ℓ of the rectangle is five times the width w"?

   A $w = 5 + \ell$

   B $w = 5\ell$

   C $\ell = 5w$

   D $\ell = 5 + w$

   E $\ell = w - 5$

3. Which statement is false?

   A $-6 > -8$

   B $3^3 > 4^2$

   C $|-6^2| = |7^2 - 13|$

   D $-121 < -141$

   E $5 \times 3^2 < 6 \times 2^3$

4. Which number has the greatest value?

   A 0.137

   B 0.46

   C 0.3

   D 0.2138

   E 0.152

5. Evaluate the expression $6 \cdot (4^2 + 6)$.

   A 30

   B 54

   C 84

   D 102

   E 132

6. Estimate the product 7.38×19.716 by rounding to the nearest one.

   A 114

   B 120

   C 140

   D 146

   E 160

7. Find the area of this figure, round your final answer to the nearest whole number of square units.

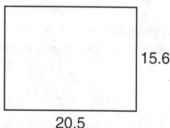

   15.6

   20.5

   A 72

   B 280

   C 319

   D 320

   E 336

8. Martin washed his vehicle exactly once every week for 2 years in a row. How many times did he wash his vehicle during this time period?

   A 24 times

   B 100 times

   C 104 times

   D 124 times

   E 144 times

Grade 9

9. A gym class had 42 students and 7 volleyballs. What was the ratio of volleyballs to students?

A 1 to 6

B 6 to 1

C 1 to 7

D 7 to 1

E 1 to 42

10. At the baseball stadium, 3 hotdogs cost $8.25. At this price, how much will 7 hotdogs cost? Which proportion could be used to solve the problem?

A $\dfrac{\$8.25}{3} = \dfrac{7}{x}$

B $\dfrac{\$8.25}{3} = \dfrac{x}{7}$

C $\dfrac{\$8.25}{x} = \dfrac{7}{3}$

D $\dfrac{\$8.25}{3} = \dfrac{x}{10}$

E $\dfrac{\$8.25}{10} = \dfrac{x}{7}$

11. What is the scale factor for the two similar figures shown below, in simplest form?

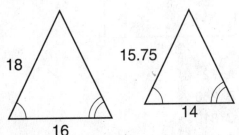

A 16 to 14

B 9 to 7

C 15.75 to 16

D 8 to 7

E 9 to 8

12. A ball player makes 538 field goals out of 934 attempts. What percent of his field goals did he make?

A 28.2%

B 34.6%

C 52.2%

D 57.6%

E 173.6%

13. What is 32% of 125?

A 20

B 32

C 40

D 42

E 44

14. The number 8 is a factor of which of the following numbers?

A 21

B 57

C 81

D 92

E 104

15. What is the greatest common factor of 36 and 42?

A 3

B 4

C 6

D 9

E 252

16. How much tax is charged to purchase 5 lamps at $65 each, if the sales tax rate is 8%?

A $1.04

B $5.20

C $10.40

D $26.00

E $52.00

Grade 9

Name _____ Date _____ Class _____

Measurement

17. What is the length of the figure shown?

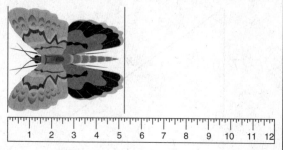

A 5 inches

B 5.125 inches

C $5\frac{1}{4}$ inches

D 5.375 inches

E $5\frac{1}{2}$ inches

18. What is the circumference of a circle with radius 8 inches? (Use 3.14 for π.)

A 12.56 inches

B 25.12 inches

C 50.24 inches

D 100.48 inches

E 200.96 inches

19. What is the perimeter of the figure?

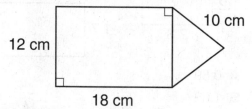

A 68 cm

B 74 cm

C 78 cm

D 80 cm

E 112 cm

20. What is the volume, in cubic inches, of a cylinder with a radius of 4 inches and a height of 12 inches? (Use 3.14 for π.)

A 150.72

B 301.44

C 602.88

D 1205.76

E 1808.64

21. What is the surface area of the prism?

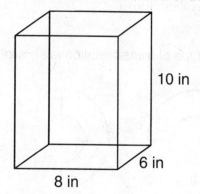

A 168 square inches

B 216 square inches

C 240 square inches

D 376 square inches

E 480 square inches

22. Mr. and Mrs. Kingston are installing fence around a rectangular piece of property. The length of the property is 780 yards and the width is 520 yards. How many yards of fencing are needed?

A 1300 yards   D 2860 yards

B 1560 yards   E 405,600 yards

C 2600 yards

23. Joel canoes 6.2 miles in 85 minutes. What is his rate in miles per hour?

A 13.4 mph   D 1.4 mph

B 6.2 mph   E 0.7 mph

C 4.3 mph

Grade 9

Geometry

24. What is the transformation from figure 1 to figure 2?

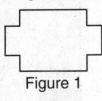

Figure 1

Figure 2

- **A** slide, turn
- **B** flip, turn
- **C** turn
- **D** slide, slide, turn
- **E** slide, flip

25. Which type of transformation is shown below?

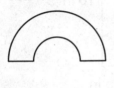

- **A** translation
- **B** rotation
- **C** reflection
- **D** glide reflection
- **E** dilation

26. The figures shown are similar. What is the value of n?

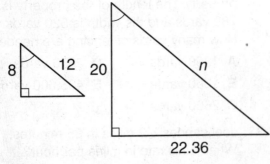

- **A** 4.8
- **B** 30
- **C** 32
- **D** 48
- **E** 56

27. What is the value of x?

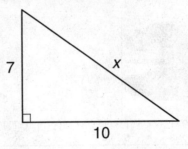

- **A** 3
- **B** $\sqrt{149}$
- **C** $\sqrt{170}$
- **D** 17
- **E** 34

28. Which lengths below would form a right triangle?

- **A** 1, 3, 5
- **B** 1.2, 2.4, 6.1
- **C** 6, 9, 10
- **D** $2\sqrt{5}$, 4, 6
- **E** $3\sqrt{2}$, $\sqrt{2}$, $\sqrt{2}$

29. The two rectangles are similar. What is the measure of the unknown side? Round your answer to the nearest hundredth.

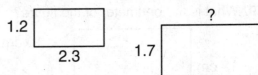

- **A** 0.89
- **B** 1.62
- **C** 2.80
- **D** 3.26
- **E** 5.94

30. A right triangle has one leg that is 5 inches longer than the other leg. The hypotenuse is 25 inches long. Find the length of the longer leg.

- **A** 10 in.
- **B** 13 in.
- **C** 15 in.
- **D** 17 in.
- **E** 20 in.

Grade 9

Name _____ Date _____ Class _____

31. A parallelogram has an area of 180 square inches and the height is 12 inches. What is the length of the base?

A 6 inches

B 15 inches

C 18 inches

D 90 inches

E 2160 inches

32. What is the slope of the line?

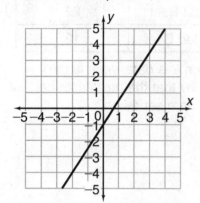

A −1

B 1

C $\frac{2}{3}$

D $\frac{3}{2}$

E 2

33. What is the midpoint of $(-8, -4)$ and $(14, 8)$?

A $(-3, -2)$

B $(3, 2)$

C $(-11, -6)$

D $(11, 6)$

E $(6, 4)$

34. What is the slope of a line parallel to $4x - 3y = 12$?

A −4

B $-\frac{3}{4}$

C $\frac{1}{4}$

D $\frac{3}{4}$

E $\frac{4}{3}$

Data Analysis, Statistics, and Probability

Use the data in the table for Questions 35 to 38.

The data shows the top five ranked men's tennis players, their total earned points, and number of tournaments played.

Player	Points Earned	Number of Tournaments Played
Roddick	3085	20
Federer	6225	18
Hewitt	2490	17
Agassi	2275	17
Nadal	4765	24

35. Who has earned the greatest number of points?

A Roddick

B Federer

C Hewitt

D Agassi

E Nadal

36. What is the average number of points Agassi has scored in a tournament?

A 133.8 points

B 198.5 points

C 146.5 points

D 154.3 points

E 345.8 points

37. If using the data in the table to construct a bar graph based on the number of tournaments played, the height of which two players would be the same?

A Federer and Hewitt

B Nadal and Hewitt

C Agassi and Roddick

D Roddick and Nadal

E Hewitt and Agassi

38. What is the median of the points earned by the top five players?

A 3085

B 6225

C 2490

D 2275

E 4765

Grade 9

Use the box and whisker plot for Questions 39 and 40.

The Humane Society is tracking the number of puppies born in a litter in a given year.

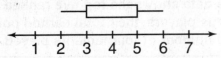

39. What is the first quartile in the distribution of data?

  A 2

  B 3

  C 4

  D 5

  E 6

40. What is the difference between the median and the third quartile?

  A 1

  B 2

  C 3

  D 4

  E 5

41. The caterer for Joan and Matt's wedding has given them a choice of 3 different kinds of meat, 3 kinds of potatoes, 4 vegetables, and 2 soups. How many complete dinners are available to choose from?

  A 12

  B 18

  C 24

  D 36

  E 72

42. Regita collected data on the kind of sandwiches her customers order at her deli shop. The table shows her findings.

Sandwich	Number of Customers
Ham	26
Turkey	25
Chicken	15
Roast Beef	22
Corned Beef	12

Based on this data, what is the probability that the next customer will order a ham sandwich?

  A $\dfrac{1}{100}$

  B $\dfrac{21}{100}$

  C $\dfrac{13}{50}$

  D $\dfrac{1}{2}$

  E $\dfrac{67}{100}$

43. Which of the following is a possible sample space for this experiment?

  A {1, 2, 3, 4, 5}

  B {Roy, Becky, George}

  C {head, tails}

  D {Roy, Yolanda, Paul, George, Becky, Walt}

  E {1, 4, heads, tails}

Grade 9

44. What is the probability of spinning a 1 then a 2 on the next spin?

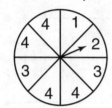

A $\frac{1}{4}$

B $\frac{1}{64}$

C $\frac{1}{16}$

D $\frac{1}{2}$

E $\frac{1}{8}$

45. Several numbers and letters are put on cards. Jamil hangs them on a wall as shown in the picture and randomly throws a dart at one of the cards.

5	A	3	S
B	Z	8	M
N	7	2	U
E	Q	P	12

What is the probability of landing on a vowel or an odd number?

A $\frac{1}{5}$

B $\frac{5}{16}$

C $\frac{3}{8}$

D $\frac{7}{16}$

E $\frac{9}{14}$

Algebra and Functions

46. What is the common ratio for the given sequence?
$4, -6, 9, -13.5, 20.25$

A -1.5

B 1.5

C 2

D -2

E 3

47. What is the missing number in the table?

x	y
−2	3
0	7
2	11
4	15
6	?

A 12   D 21

B 17   E 28

C 19

48. Miguel made a display of DVD's at a movie rental store. One DVD was in the first row, and the other rows each had two more DVD's than the row before it. How many DVD's does Miguel have if he has nine rows?

A 19

B 27

C 36

D 81

E 121

49. Solve for x.
$6x + 13 = 49$

A -6

B 5

C 6

D $10\frac{1}{3}$

E 12

Grade 9

50. What are the coordinates of point C?

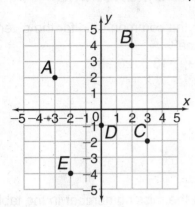

   A $(-3, 2)$

   B $(0, -1)$

   C $(2, 4)$

   D $(3, -2)$

   E $(-2, -4)$

51. Plot and connect the given points in order, and determine the perimeter of the figure.

$(-4, 2), (-2, 2), (-2, 4), (1, 4), (1, 3),$
$(3, 3), (3, -4), (-4, -4)$

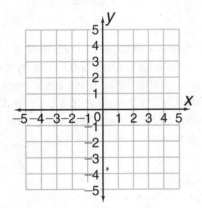

   A 24   **D** 32

   B 26   **E** 36

   C 30

52. Which of the following is the equation of $y = -4x - 3$ shifted 4 units down?

   A $y = -4x + 1$

   B $y = -4x - 7$

   C $y = 4x - 7$

   D $y = 4x + 1$

   E $y = -4x - 8$

53. Which graph represents the solution of $2x + 12 > 16$?

A
```
<-+--+--+--+--+--+--+--+--●--+--+--+->
 -5 -4 -3 -2 -1  0  1  2  3  4  5
```

B
```
<-+--+--+--+--+--+--+--+--●━━+━━+━━+━>
 -5 -4 -3 -2 -1  0  1  2  3  4  5
```

C
```
<-×━━+━━+━━+━━+━━+━━+━━+━━●--+--+--+->
 -5 -4 -3 -2 -1  0  1  2  3  4  5
```

D
```
<-+--+--+--+--+--+--+--+--⊕━━+━━+━━×━>
 -5 -4 -3 -2 -1  0  1  2  3  4  5
```

E
```
<-×━━+━━+━━+━━+━━+━━+━━+━━⊕--+--+--+->
 -5 -4 -3 -2 -1  0  1  2  3  4  5
```

54. Simplify: $\dfrac{4x^2\sqrt{6^2 + 8^2}}{8x}$

   A $5x$

   B $10x$

   C $0.5x$

   D $30x$

   E $20x$

55. Which expression is equivalent to $-3(b^2 - 4b)$?

   A $3b^2 + 4b$

   B $-3b^2 - 12b$

   C $9b^2 + 12b$

   D $-3b^2 + 4b$

   E $12b - 3b^2$

56. Which expression is the completely factored form of $x^3 - x^2 - 26x + 30$?

   A $(x + 5)(x^2 - 4x + 6)$

   B $(x - 5)(x^2 + 4x - 6)$

   C $(x + 5)(x^2 + 4x - 6)$

   D $(x - 5)(x^2 - 4x + 6)$

   E $(x + 5)(x^2 - 4x - 6)$

Grade 9

Name _____   Date _____ Class _____

57. Which graph is the solution to the system of equations?

$y = x - 2$

$2x - 3y = -9$

A

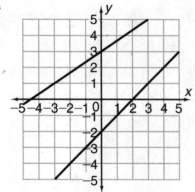

B

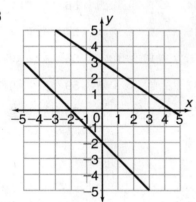

C

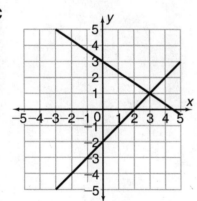

D

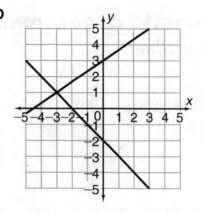

E

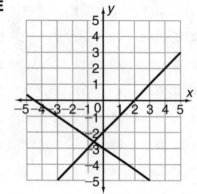

58. What is the x-value in the following system?

$3x - 2y = 1$

$x + 2y = 11$

A 0   **D** 3

B 1   **E** 4

C 2

59. What is the value of m in the matrix?

$$\begin{bmatrix} 6 & -2 \\ m & 0 \end{bmatrix} = \begin{bmatrix} n & -2 \\ -4 & 0 \end{bmatrix}$$

A $m = -4$ **D** $m = 4$

B $m = -2$ **E** $m = 6$

C $m = 0$

60. What is the value of y in the matrix below?

$$\begin{bmatrix} -12 & 10 & -11 \\ 10 & 11 & 12 \end{bmatrix} + \begin{bmatrix} 8 & 6 & -5 \\ -8 & 6 & 5 \end{bmatrix} = \begin{bmatrix} 4 & 16 & -16 \\ 2 & y & 17 \end{bmatrix}$$

A -16 **D** 16

B -6 **E** 17

C 4

Grade 9

Name_____ Date _____ Class _____

Diagnostic Assessment
Answer Sheet

1 Ⓐ Ⓑ Ⓒ Ⓓ Ⓔ	13 Ⓐ Ⓑ Ⓒ Ⓓ Ⓔ	25 Ⓐ Ⓑ Ⓒ Ⓓ Ⓔ	37 Ⓐ Ⓑ Ⓒ Ⓓ Ⓔ	49 Ⓐ Ⓑ Ⓒ Ⓓ Ⓔ
2 Ⓐ Ⓑ Ⓒ Ⓓ Ⓔ	14 Ⓐ Ⓑ Ⓒ Ⓓ Ⓔ	26 Ⓐ Ⓑ Ⓒ Ⓓ Ⓔ	38 Ⓐ Ⓑ Ⓒ Ⓓ Ⓔ	50 Ⓐ Ⓑ Ⓒ Ⓓ Ⓔ
3 Ⓐ Ⓑ Ⓒ Ⓓ Ⓔ	15 Ⓐ Ⓑ Ⓒ Ⓓ Ⓔ	27 Ⓐ Ⓑ Ⓒ Ⓓ Ⓔ	39 Ⓐ Ⓑ Ⓒ Ⓓ Ⓔ	51 Ⓐ Ⓑ Ⓒ Ⓓ Ⓔ
4 Ⓐ Ⓑ Ⓒ Ⓓ Ⓔ	16 Ⓐ Ⓑ Ⓒ Ⓓ Ⓔ	28 Ⓐ Ⓑ Ⓒ Ⓓ Ⓔ	40 Ⓐ Ⓑ Ⓒ Ⓓ Ⓔ	52 Ⓐ Ⓑ Ⓒ Ⓓ Ⓔ
5 Ⓐ Ⓑ Ⓒ Ⓓ Ⓔ	17 Ⓐ Ⓑ Ⓒ Ⓓ Ⓔ	29 Ⓐ Ⓑ Ⓒ Ⓓ Ⓔ	41 Ⓐ Ⓑ Ⓒ Ⓓ Ⓔ	53 Ⓐ Ⓑ Ⓒ Ⓓ Ⓔ
6 Ⓐ Ⓑ Ⓒ Ⓓ Ⓔ	18 Ⓐ Ⓑ Ⓒ Ⓓ Ⓔ	30 Ⓐ Ⓑ Ⓒ Ⓓ Ⓔ	42 Ⓐ Ⓑ Ⓒ Ⓓ Ⓔ	54 Ⓐ Ⓑ Ⓒ Ⓓ Ⓔ
7 Ⓐ Ⓑ Ⓒ Ⓓ Ⓔ	19 Ⓐ Ⓑ Ⓒ Ⓓ Ⓔ	31 Ⓐ Ⓑ Ⓒ Ⓓ Ⓔ	43 Ⓐ Ⓑ Ⓒ Ⓓ Ⓔ	55 Ⓐ Ⓑ Ⓒ Ⓓ Ⓔ
8 Ⓐ Ⓑ Ⓒ Ⓓ Ⓔ	20 Ⓐ Ⓑ Ⓒ Ⓓ Ⓔ	32 Ⓐ Ⓑ Ⓒ Ⓓ Ⓔ	44 Ⓐ Ⓑ Ⓒ Ⓓ Ⓔ	56 Ⓐ Ⓑ Ⓒ Ⓓ Ⓔ
9 Ⓐ Ⓑ Ⓒ Ⓓ Ⓔ	21 Ⓐ Ⓑ Ⓒ Ⓓ Ⓔ	33 Ⓐ Ⓑ Ⓒ Ⓓ Ⓔ	45 Ⓐ Ⓑ Ⓒ Ⓓ Ⓔ	57 Ⓐ Ⓑ Ⓒ Ⓓ Ⓔ
10 Ⓐ Ⓑ Ⓒ Ⓓ Ⓔ	22 Ⓐ Ⓑ Ⓒ Ⓓ Ⓔ	34 Ⓐ Ⓑ Ⓒ Ⓓ Ⓔ	46 Ⓐ Ⓑ Ⓒ Ⓓ Ⓔ	58 Ⓐ Ⓑ Ⓒ Ⓓ Ⓔ
11 Ⓐ Ⓑ Ⓒ Ⓓ Ⓔ	23 Ⓐ Ⓑ Ⓒ Ⓓ Ⓔ	35 Ⓐ Ⓑ Ⓒ Ⓓ Ⓔ	47 Ⓐ Ⓑ Ⓒ Ⓓ Ⓔ	59 Ⓐ Ⓑ Ⓒ Ⓓ Ⓔ
12 Ⓐ Ⓑ Ⓒ Ⓓ Ⓔ	24 Ⓐ Ⓑ Ⓒ Ⓓ Ⓔ	36 Ⓐ Ⓑ Ⓒ Ⓓ Ⓔ	48 Ⓐ Ⓑ Ⓒ Ⓓ Ⓔ	60 Ⓐ Ⓑ Ⓒ Ⓓ Ⓔ

Grade 9

Name _____ Date _____ Class _____

Test Preparation Practice
Number Properties and Operations

12.1.1.d Write, rename, represent, or compare real numbers (e.g., pi, square root of 2, numerical relationships using number lines, models, or diagrams).

Solve each problem, and circle the letter of the best answer.

1. Which number has the greatest absolute value?

   A $\dfrac{-120}{2}$

   B -59.9

   C $\dfrac{206}{4}$

   D 59.75

   E $\dfrac{160}{4}$

2. What is another way to write 3^4?

   A 12

   B 24

   C 27

   D 81

   E 243

3. Which of the following is the greatest integer less than 54,000 that can be written using all of the digits from 1 to 5?

   A 54,321

   B 53,241

   C 54,312

   D 53,421

   E 53,241

4. Which of the following fractions is equivalent to 0.4375?

   A $\dfrac{1}{31}$ **D** $\dfrac{5}{6}$

   B $\dfrac{7}{16}$ **E** $\dfrac{7}{9}$

   C $\dfrac{5}{8}$

5. Which shows the correct order of numbers from least to greatest?

   A $\sqrt{18}, 4.9, \dfrac{21}{4}, \dfrac{31}{8}$

   B $4.9, \dfrac{21}{4}, \sqrt{18}, \dfrac{31}{8}$

   C $\dfrac{21}{4}, \dfrac{31}{8}, 4.9, \sqrt{18}$

   D $\dfrac{31}{8}, \sqrt{18}, 4.9, \dfrac{21}{4}$

   E $\sqrt{18}, 4.9, \dfrac{31}{8}, \dfrac{21}{4}$

6. Which of these values is not an integer?

   A -2

   B 0

   C $\sqrt{36}$

   D $\dfrac{16}{4}$

   E $-\dfrac{1}{2}$

7. The number one-half can be written in each of the following ways except:

   A $\dfrac{4}{2}$ **D** $\dfrac{1}{2}$

   B $\dfrac{50}{100}$ **E** 0.5

   C $2 \div 4$

Grade 9

8. Which statement is false?

A $-3.02 > -3.2$

B $\left| -\dfrac{6\pi}{2} \right| = 3\pi$

C $4^2 = 2^4$

D $\left| 13^2 - 13 \right| > \left| 13 - 13^2 \right|$

E $-5^3 = (-5)^3$

9. Which of the number sentences below is false?

A $-4 \le -2$

B $-8 > -12$

C $-6 > -10$

D $-7 < 0$

E $-5 < -8$

10. Rewrite $\dfrac{3^9}{3^3}$.

A 1^3

B 1^6

C 3^3

D 3^6

E 3^{12}

11. The outdoor temperature is negative six degrees Fahrenheit. Which one of the following corresponds to this statement?

A $-8\ °F$

B $-6\ °F$

C $-2\ °F$

D $6\ °F$

E $16\ °F$

12. How would you write these elevations from lowest point to highest point?

Location	Elevation
Death Valley	−86 m
Mt. McKinley	6194 m
Turpan Pendi	−154 m
Mt. Everest	8850 m

A Turpan Pendi, Death Valley, Mt. McKinley, Mt. Everest

B Death Valley, Turpan Pendi, Mt. Everest, Mt. McKinley

C Death Valley, Mt. McKinley, Turpan Pendi, Mt. Everest

D Mt. Everest, Mt. McKinley, Turpan Pendi, Death Valley

E Turpan Pendi, Mt. Everest, Mt. McKinley, Death Valley

13. What is eight hundred nine thousand thirty written in standard form?

A 809

B 890,030

C 89,030

D 809,300

E 809,030

14. The square root of one hundred ninety-six can be represented by all of the following values except:

A $\sqrt{14 \cdot 14}$

B $4(7)$

C $\sqrt{196}$

D 14

E $\sqrt{4 \cdot 49}$

Test Preparation Practice
Number Properties and Operations

12.1.1.f Represent very large or very small numbers using scientific notation in meaningful contexts.

Solve each problem, and circle the letter of the best answer.

1. The age of our solar system is about 5.1×10^9 years. Which is another way to express this age?

   A 510,000

   B 510,000,000

   C 5,100,000,000

   D 510,000,000,000

   E 510,000,000,000,000

2. Which of these numbers is written in scientific notation?

   A 92×10^{15}

   B 9.2×10^8

   C 920^8

   D $9.2 + 10^{33}$

   E 0.92×10^6

3. The mass of a proton is 0.0000000000000000000001672 milligrams. Express this mass in scientific notation.

   A 1672×10^{-15} mg

   B 16.72×10^{-20} mg

   C 1.672×10^{-21} mg

   D 1.672×10^{-22} mg

   E 1.672×10^{-23} mg

4. If the number of molecules in one mole of a compound is 8.03×10^9, then the number of molecules in 10,000 moles of the same compound is:

   A 8.03×10^5

   B 8.03×10^{13}

   C 8.03×10^{23}

   D 8.03×10^{33}

   E 8.03×10^{39}

5. A laboratory technician is looking through a microscope that has magnified an organism by 1000. If a bacteria is 3.1×10^{-4} millimeters, how large does it appear under the microscope?

   A 3.1×10^{-3} mm

   B 3.1×10^{-2} mm

   C 3.1×10^{-1} mm

   D 3.1×10^1 mm

   E 3.1×10^2 mm

6. If the speed of light is 3.00×10^8 meters per second, how far would a beam of light travel in 3000 seconds?

   A 1.0×10^5 m

   B 1.0×10^8 m

   C 9.0×10^5 m

   D 9.0×10^8 m

   E 9.0×10^{11} m

Grade 9

7. A particle travels 1×10^8 centimeters per second in a straight line for 5×10^{-6} second. How far has the particle traveled?

   A 5×10^{14} cm

   B 5×10^2 cm

   C 2.0×10^2 cm

   D 2.0×10^{-13} cm

   E 5×10^{-48} cm

8. A runner burns about 350 calories per hour. If there are 16,000 runners in a marathon, about how many calories would be burned by all of the runners in one hour?

   A 4.5×10^2 cal

   B 5.6×10^4 cal

   C 5.6×10^5 cal

   D 5.6×10^6 cal

   E 5.6×10^{12} cal

9. A gigabyte is a measure of a computer's storage capacity. One gigabyte holds about one billion bytes of information. If a company's computer network contains 3200 gigabytes of memory. How many bytes are in the network?

   A 3.2×10^9

   B 3.2×10^{10}

   C 3.2×10^{11}

   D 3.2×10^{12}

   E 3.2×10^{13}

10. The Earth is about 93 million miles from the sun. Write the underlined number in scientific notation.

   A 93×10^6

   B 93×10^7

   C 9.3×10^6

   D 9.3×10^7

   E 9.3×10^8

11. Light travels from the Sun to Earth at a rate of about 300,000 kilometers per second. If a light beam takes 492 seconds to reach the Earth. Find the distance from the Sun to Earth.

   A 1.64×10^7 kilometers

   B 1.64×10^8 kilometers

   C 1.92×10^8 kilometers

   D 1.476×10^7 kilometers

   E 1.476×10^8 kilometers

12. On Mother's Day, 2005, the United States Census Bureau estimated there were approximately 62.5 million women in the United States who were mothers. If each mother received one $2.00 greeting card, approximately how much money would the greeting card industry earn on Mother's Day?

   A $\$1.25 \times 10^8$

   B $\$6.25 \times 10^7$

   C $\$3.15 \times 10^7$

   D $\$1.25 \times 10^6$

   E $\$3.15 \times 10^6$

Grade 9

Name _____ Date _____ Class _____

Test Preparation Practice

Number Properties and Operations

12.1.2.a Establish or apply benchmarks for real numbers in contexts.

Solve each problem, and circle the letter of the best answer.

1. A tube of antibiotic cream contains 38.375 grams of cream. What is this amount rounded to the nearest tenth of a gram?

    A 38.0 g

    B 38.3 g

    C 38.4 g

    D 38.48 g

    E 40 g

2. What is the estimated product when 167 and 917 are rounded to the nearest hundred and then multiplied?

    A 80,000

    B 90,000

    C 100,000

    D 180,000

    E 200,000

3. Lisa has a picture frame that measures 3.5 feet by 4.25 feet. She needs to purchase a piece of glass to keep her picture in the frame. What is the area, rounded to the nearest hundredth, of the piece of glass she should buy?

    A 14.0 square feet

    B 14.87 square feet

    C 14.88 square feet

    D 15.0 square feet

    E 15.13 square feet

4. Keisha bought the following items at the grocery store. Estimate the total, rounded to the nearest whole dollar.

Item	Cost
lettuce	$0.79
raspberries	$2.99
milk	$3.29
cereal	$4.49
rice	$0.99
pancake mix	$2.19

    A $10

    B $12

    C $13

    D $14

    E $18

5. A 16-ounce can of soup sells for $2.79. How much does one ounce of soup cost, rounded to the nearest cent?

    A $0.15

    B $0.17

    C $0.21

    D $0.45

    E $44.64

6. Maria lives in a condominium that is $56\frac{3}{5}$ meters tall. What is this number rounded to the nearest half meter?

    A 56 m

    B $56\frac{1}{2}$ m

    C 57 m

    D $57\frac{1}{2}$ m

    E 58 m

Grade 9

Name _____   Date _____ Class _____

7. The average daily volume of a particular stock being traded on Wall Street is 8.259 Mil, what is this number rounded to the nearest hundredth?

A 8.0 Mil

B 8.2 Mil

C 8.25 Mil

D 8.26 Mil

E 8.3 Mil

8. Santos computed that with his last tank of gas he averaged 26.64 miles per gallon. What is this rate rounded to the nearest one?

A 25 mph

B 26 mph

C 26.6 mph

D 27 mph

E 27.6 mph

9. The school cafeteria had $14\frac{2}{3}$ gallons of applesauce left after serving lunch. What is this number rounded to the nearest whole gallon?

A 14 gallons

B $14\frac{1}{2}$ gallons

C 15 gallons

D $15\frac{1}{2}$ gallons

E 16 gallons

10. In 2001, the population of the United States was 281,421,906. What is this number rounded to the nearest ten million?

A 280,000,000

B 281,000,000

C 281,400,000

D 281,500,000

E 290,000,000

11. What is the length of the bee rounded to the nearest quarter inch?

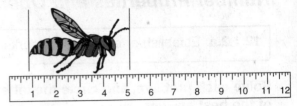

A $5\frac{1}{2}$ in.

B $5\frac{1}{4}$ in.

C 5 in.

D $4\frac{1}{2}$ in.

E $4\frac{1}{4}$ in.

12. The managers of the state fair estimated that the attendance on a particular day was 16,000. Assuming that the customary rules for rounding have been used, which of the following could be the exact number of people who attended the fair?

A 15,099

B 15,304

C 15,444

D 16,333

E 16,566

13. The diameter *d* of a circle is 12 inches. The circumference *C* of a circle is $C = \pi d$. Estimate the circumference of the circle.

A 4 in.

B 12 in.

C 18 in.

D 36 in.

E 48 in.

Grade 9

Name _____ Date _____ Class _____

Test Preparation Practice
Number Properties and Operations

12.1.3.a Perform computations with real numbers including common irrational numbers or the absolute value of numbers.

Solve each problem, and circle the letter of the best answer.

1. The Jarvis family consists of two adults and four children. An adult admission to a local amusement park is $49 and a child admission is $28. The Jarvis' have a coupon for $10 off each admission for up to four admissions. How much will it cost the family to enter the park?

   A $150

   B $170

   C $210

   D $212

   E $234

2. Evaluate the expression $-8.2(6 + 5.2)$.

   A 11.2

   B 3

   C -6.56

   D -44

   E -91.84

3. Evaluate the expression $\frac{1}{2} + \frac{1}{4}(20)$.

   A 2

   B 2.5

   C 4.5

   D 5.5

   E 10

Use the table for Questions 4 and 5.

Grove Hill Campground Rates		
	Large	Big Rig
E	$23	$28
E, W	$25	$30
E, W, S	$30	$35
E = electric, W = water, S = sewer		
Stay 7 nights, the 8th night is free.		

4. The Ortez family went camping. They planned to stay 7 nights at a large site with electricity and water. After one night, they moved to a big rig site with electric, water and sewer. They stayed the remainder of their vacation at this site. How much did they spend in all for their stay at the campground?

   A $175

   B $205

   C $235

   D $240

   E $245

5. The Schmidt family is making reservations at the Grove Hill campground for 10 nights. They have reserved a big rig site with electric and water. How much will the Schmidt family spend for their stay at the campground?

   A $225

   B $270

   C $280

   D $300

   E $315

Grade 9

6. The Oakdale school has 1326 students and 125 teachers. On Monday 156 students and 8 teachers were absent because of the flu. How many students and teachers were present on Monday?

   A 1162

   B 1170

   C 1287

   D 1326

   E 1451

7. A computer company needs to build 4500 new computers in 6 weeks. It takes 7 hours to build one computer. Employees work 35 hours per week. How many people does the company need to hire in order to build the computers?

   A 19

   B 129

   C 140

   D 150

   E 191

8. Evaluate the expression:

   $|18 - 29| - |-12|$

   A 23

   B 1

   C 21

   D -1

   E -3

9. Cody bought tickets for the local county fair. The table shows the prices of single tickets.

Ticket Prices	
Child	$2.25
Adult	$6.00
Senior	$3.25

   Cody bought 3 child tickets, two adult tickets and one senior ticket. How much money did Cody spend on tickets?

   A $10.00

   B $16.00

   C $18.75

   D $22.00

   E $24.00

10. Mitchell earned $178.50 last week at his job at the deli. His hourly wage is $7.00. How many hours did Mitchell work last week?

    A 3.9 hours

    B 22 hours

    C 25 hours

    D 25.5 hours

    E 27.5 hours

11. Mrs. Diamond is giving each of her 29 students a holiday pencil. The pencils come in packages of 8. Each package costs $1.50. How much will Mrs. Diamond spend on pencils for her students?

    A $4.50

    B $6.00

    C $9.00

    D $9.50

    E $12.00

Grade 9

Test Preparation Practice
Number Properties and Operations

12.1.3.g Solve application problems involving numbers, including rational and common irrationals, using exact answers or estimates as appropriate.

Solve each problem, and circle the letter of the best answer.

1. During one month, the cafeterias at Southwest State University prepared 23,295 meals. The next month it prepared 32,188 meals. The following month, 54,848 meals were prepared. Which choice is the best estimate for the total number of meals prepared during the three months?

   A 100,000

   B 109,000

   C 110,000

   D 112,000

   E 115,000

2. A school district purchased 20,000 fluorescent light bulbs for the 18 schools in the district. A total of 62 bulbs were replaced in the gymnasium of one high school. Of these, 4 were defective. What is the best estimate of the total number of defective bulbs purchased by the district?

   A 10

   B 200

   C 400

   D 800

   E 1300

3. Victor walks $\frac{3}{5}$ of a mile to school. After school he walks $\frac{2}{3}$ of a mile to the library and then $\frac{5}{8}$ of a mile home. Which is the best estimate for the total distance Victor walks per day?

   A 1 mile

   B $1\frac{1}{2}$ miles

   C 2 miles

   D 3 miles

   E 5 miles

4. Susan and Adrian go to Martha's Café for lunch. Susan orders a turkey sandwich, salad, and juice. Adrian orders a cheese sandwich, soup, and juice. They want to estimate the bill to figure out the tip. What is the best estimate of the total bill before taxes and tip?

Martha's Cafe	
Turkey sandwich	$4.75
Cheese sandwich	$2.25
Tuna sandwich	$2.90
Salad	$2.75
Soup	$2.50
Juice	$1.35
Milk	$0.75

   A $3

   B $5

   C $10

   D $12

   E $15

Grade 9

5. To make a large pot of soup, a chef used $2\frac{1}{4}$ cups of carrots, $1\frac{1}{3}$ cups of corn, 6 cups of potatoes, $2\frac{3}{4}$ cups of peas, and $\frac{3}{4}$ cup of celery. What is the best estimate of the total amount of vegetables used in the soup?

A 10 cups

B 12 cups

C 13 cups

D 16 cups

E 20 cups

6. Gian spends between $9 and $15 for lunch per week. What is the best estimate for the total amount that he spent on lunches for the past 4 weeks?

A less than $30

B between $40 and $45

C between $40 and $50

D between $40 and $60

E more than $60

7. Mario is a painter. He estimates that a fence he is painting has a length between 100 and 130 feet and a width between 30 to 60 feet. Based on Mario's estimates, which is the most reasonable estimate of the perimeter of the fence?

A less than 250 feet

B between 260 and 380 feet

C between 460 and 500 feet

D between 500 and 600 feet

E over 900 feet

8. Roan needs to cover a circular area that has a diameter of 12 feet. What is the best estimate of this area?

A 18 ft^2

B 36 ft^2

C 72 ft^2

D 108 ft^2

E 144 ft^2

9. Leo has 2 jobs during the summer. He works 22 hours per week at the first job for $11 per hour. He works 9 hours per week at the second job for $18 per hour. Which is the best estimate of the total amount of money Leo earns from both jobs in one week?

A $40

B $180

C $220

D $400

E $600

10. Hugo bought a new refrigerator on sale for $929. The regular price for the refrigerator was $1592. Which is the best estimate of the total amount Hugo saved?

A $200

B $500

C $700

D $800

E $1000

11. Karen pours water into 4 buckets. The buckets contain 7.84 liters, 10.24 liters, 4.84 liters and 2.23 liters. Which is the best estimate of the total amount of water in the 4 buckets?

A 10 L

B 15 L

C 20 L

D 25 L

E 35 L

12. A roofing company estimates that a new roof for Kris's house will cost $5580. If Kris pays for the roof in 18 equal payments, what is the best estimate for the amount of each payment?

A $120

B $200

C $250

D $300

E $400

Grade 9

Name _____ Date _____ Class _____

Test Preparation Practice

Number Properties and Operations

12.1.4.b Use proportions to model problems.

Solve each problem, and circle the letter of the best answer.

1. Brady knows that 8 tickets can be purchased for $68, but needs to know how many can be purchased for $102. Which proportion models this situation?

   A $\dfrac{x}{102} = \dfrac{68}{8}$

   B $\dfrac{8}{102} = \dfrac{x}{68}$

   C $\dfrac{68}{8} = \dfrac{102}{x}$

   D $\dfrac{102}{8} = \dfrac{68}{x}$

   E $\dfrac{x}{8} = \dfrac{68}{102}$

2. There are 90 calories in a $\dfrac{2}{3}$ cup serving of cereal. How many calories are in 4 cups of cereal?

   A 60

   B 240

   C 380

   D 480

   E 540

Use the information below for Questions 3 and 4.

Gwen can spray paint 850 ft² with 2 gallons of paint. How many one-gallon cans does she needs to purchase in order to spray paint a building that is 24,000 square feet.

3. Which proportion could Gwen use to determine how many one-gallon cans she needs?

   A $\dfrac{850 \text{ ft}^2}{1 \text{ gal}} = \dfrac{24{,}000 \text{ ft}^2}{x}$

   B $\dfrac{850 \text{ ft}^2}{2 \text{ gal}} = \dfrac{24{,}000 \text{ ft}^2}{x}$

   C $\dfrac{850 \text{ ft}^2}{2 \text{ gal}} = \dfrac{x}{24{,}000 \text{ ft}^2}$

   D $\dfrac{850 \text{ ft}^2}{24{,}000 \text{ ft}^2} = \dfrac{x}{2 \text{ gal}}$

   E $\dfrac{24{,}000 \text{ ft}^2}{2 \text{ gal}} = \dfrac{x}{850 \text{ ft}^2}$

4. How many one-gallon cans does Gwen need to spray paint the building?

   A 27 cans

   B 28 cans

   C 56 cans

   D 57 cans

   E 62 cans

Grade 9

5. Choose the correct proportion to find the distance across the lake, assuming the triangles are similar.

A $\dfrac{15}{60} = \dfrac{45}{d}$

B $\dfrac{60}{15} = \dfrac{45}{d}$

C $\dfrac{60}{d} = \dfrac{45}{15}$

D $\dfrac{75}{15} = \dfrac{45}{d}$

E $\dfrac{74}{45} = \dfrac{x}{d}$

6. Which of the following proportions models the statement eight is to seven as x is to forty?

A $\dfrac{8}{7} = \dfrac{40}{x}$

B $\dfrac{7}{8} = \dfrac{x}{40}$

C $\dfrac{8}{7} = \dfrac{x}{40}$

D $\dfrac{7}{x} = \dfrac{40}{8}$

E $\dfrac{x}{7} = \dfrac{8}{40}$

Use the information to answer Questions 7 and 8.

Thomas is a quality control inspector. He examined 113 cabinet doors in one hour. He found 9 of them to be defective. He anticipates that he will inspect 745 doors during the remaining 7 hours of his shift.

7. Which proportion models cabinet doors inspected to defective doors found in the first hour compared to the remainder of the shift?

A $\dfrac{x}{113 \text{ doors}} = \dfrac{745 \text{ doors}}{7 \text{ hours}}$

B $\dfrac{x}{7 \text{ hours}} = \dfrac{113 \text{ doors}}{1 \text{ hour}}$

C $\dfrac{x}{745 \text{ doors}} = \dfrac{113 \text{ doors}}{7 \text{ hours}}$

D $\dfrac{745 \text{ doors}}{113 \text{ doors}} = \dfrac{9 \text{ defective}}{x}$

E $\dfrac{113 \text{ doors}}{9 \text{ defective}} = \dfrac{745 \text{ doors}}{x}$

8. How many defective cabinet doors can Thomas expect to find during the 8-hour shift?

A 52

B 60

C 65

D 68

E 72

9. Donovan's van travels 137 miles using 4.75 gallons of gas. Which proportion models how many miles he can travel using 22.5 gallons of gas?

A $\dfrac{137 \text{ gallons}}{4.75} = \dfrac{x}{22.5 \text{ gallons}}$

B $\dfrac{137 \text{ miles}}{4.75 \text{ gallons}} = \dfrac{x}{22.5 \text{ gallons}}$

C $\dfrac{137 \text{ miles}}{22.5 \text{ gallons}} = \dfrac{x}{4.75 \text{ gallons}}$

D $\dfrac{137 \text{ miles}}{22.5 \text{ gallons}} = \dfrac{4.75 \text{ gallons}}{x}$

E $\dfrac{x}{22.5 \text{ gallons}} = \dfrac{4.75 \text{ gallons}}{137 \text{ miles}}$

Grade 9

Name _____ Date _____ Class _____

Test Preparation Practice
Number Properties and Operations

12.1.4.c Use proportional reasoning to solve problems (including rates).

Solve each problem, and circle the letter of the best answer.

1. The proportion *a* is to *b* as *c* is to *d* is the same as which of the following choices? Assume that none of the variables equal zero.

   A $\dfrac{a}{c} = \dfrac{d}{b}$

   B $ac = bd$

   C $\dfrac{a}{d} = \dfrac{c}{b}$

   D $ab = cd$

   E $ad = bc$

2. Solve for x.

   $\dfrac{x}{36} = \dfrac{9}{4}$

   A $x = 16$

   B $x = 54$

   C $x = 72$

   D $x = 81$

   E $x = 144$

3. Find the value of *x* that makes the statement true.

   $\dfrac{24}{72} = \dfrac{15}{x}$

   A $x = \dfrac{5}{24}$

   B $x = \dfrac{1}{3}$

   C $x = 3$

   D $x = \dfrac{24}{5}$

   E $x = 45$

4. Find the missing value in the proportion.

   $\dfrac{8 \text{ lb}}{\$12} = \dfrac{?}{\$18}$

   A $12

   B 12 lb

   C $12/lb

   D $16

   E 16 lb

5. A gourmet shop puts 1.5 ounces of oregano in a jar and sells it for $1.95. At that rate, how much should be charged for a 2-ounce jar? Round to the nearest cent if necessary.

   A $1.30   **D** $3.90

   B $1.46   **E** $5.85

   C $2.60

6. Murphy drove 367.5 miles and used 17.5 gallons of gas. At this rate, how much gas will he need to drive 342.3 miles? Round to the nearest hundredth of a gallon.

   A 13.85 gallons

   B 15.83 gallons

   C 16.30 gallons

   D 22.55 gallons

   E 25.20 gallons

7. The scale on a map indicates that 1.25 cm represents 40 km. Two towns on the map are 7.5 cm apart. In reality, how far apart are the towns?

   A 200 km

   B 240 km

   C 250 km

   D 280 km

   E 300 km

Grade 9

Name _____ Date _____ Class _____

8. Solve for x

$$\frac{3}{7} = \frac{x-4}{2}$$

A $x = -3\frac{1}{7}$

B $x = \frac{7}{34}$

C $x = \frac{6}{7}$

D $x = 4\frac{6}{7}$

E $x = 5\frac{1}{6}$

9. Solve for x.

$$\frac{2+x}{8} = \frac{x-5}{7}$$

A $x = -54$

B $x = -26$

C $x = 3\frac{3}{5}$

D $x = 26$

E $x = 54$

10. Solve for x.

$$\frac{3\frac{3}{5}}{4\frac{1}{2}} = \frac{x}{1\frac{1}{2}}$$

A $x = \frac{11}{15}$

B $x = \frac{5}{6}$

C $x = 1\frac{1}{15}$

D $x = 1\frac{1}{5}$

E $x = 1\frac{4}{11}$

11. If 1000 ft^2 of surface can be covered with two gallons of paint, how much surface can be covered by one quart of paint?

A 80 ft^2

B 125 ft^2

C 250 ft^2

D 500 ft^2

E 525 ft^2

12. Triangle ABC is similar to triangle DEF. $AB = 7$, $BC = 11$ and $CA = 14$. Find FD if $DE = 13$.

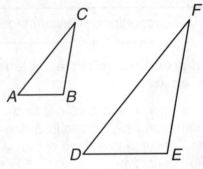

A $7\frac{7}{13}$

B $16\frac{6}{11}$

C $20\frac{3}{7}$

D 26

E $31\frac{5}{13}$

13. What is the unit rate for postcards if 12 postcards cost $4?

A 3 cards per dollar

B $0.33 per dozen

C $0.25 per dozen

D $0.33 per card

E $4 per card

14. Which model can be used to convert 50 mph to feet per second?

A 50 mph $\times \dfrac{1\ mi}{5280\ ft} \times \dfrac{1\ hr}{60\ min}$

B 50 mph $\times \dfrac{5280\ ft}{1\ mi} \times \dfrac{1\ hr}{3600\ s}$

C 50 mph $\times \dfrac{5280\ ft}{1\ mi} \times \dfrac{1\ hr}{360\ s}$

D 50 mph $\times \dfrac{5280\ ft}{1\ mi} \times \dfrac{1\ hr}{60\ s}$

E 50 mph $\times \dfrac{1\ mi}{5280\ ft} \times \dfrac{60\ min}{1\ hr}$

Grade 9

Name _____ Date _____ Class _____

Test Preparation Practice

Number Properties and Operations

12.1.4.d Solve problems involving percentages (including percent increase and decrease, interest rates, tax, discount, tips, or part/whole relationships).

Solve each problem, and circle the letter of the best answer.

1. What is 30% of 5860?

A 195.3

B 1758

C 1953.3

D 4102

E 175,800

2. Collin spends 12% of his monthly income on his car. He spends $310 a month on his car. What is his monthly income?

A $387.00

B $757.39

C $2362.67

D $2583.33

E $3720.00

3. The Silverton's are selling their house through a real-estate broker who charges a 7% sales commission. If they wish to have $132,000 after the 7% commission is subtracted, what must the selling price of the house be?

A $125,000

B $141,240

C $141,935

D $142,940

E $145,620

4. Gina pays the sale price of $270 for a computer printer that was originally $400. What percent discount is this?

A 1.48%

B 13%

C 30%

D 32.5%

E 67.5%

5. Kim compares prices of two shirts. The short-sleeved one costs $14 and the long-sleeved one costs $22. The sales tax rate is 6.5%. How much more will the long-sleeved shirt cost after tax?

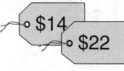

A $8.00

B $8.52

C $8.65

D $9.03

E $9.43

6. Five percent of what number is 235?

A 11.75

B 47

C 470

D 1175

E 4700

7. Amber buys an antique table for $55 at an auction and later sells it in her antique shop for $135. What is her percent profit?

A 145.5%

B 120%

C 80%

D 40.7%

E 2.5%

Grade 9

Name _____ Date _____ Class _____

8. On Wednesday 288 pints of blood are collected at a blood drive. The next day only 252 pints of blood are collected. What is the percent decrease?

A $\frac{1}{8}$%

B 4%

C $8\frac{1}{2}$%

D $12\frac{1}{2}$%

E 36%

9. On a math test, Clarence answered 84% of the questions correctly. If he answered 63 questions correctly, how many questions were on the test?

A 63

B 72

C 75

D 78

E 84

10. Bart can earn 4% interest on his money when invested properly. How much money must he invest if he wants to earn $3500 in interest per year?

A $8750

B $14,000

C $87,500

D $114,286

E $140,000

11. Lynnette sells automobile insurance. Her commission rate on all sales is 6.4%. If her sales this month were $35,672, how much commission did she earn?

A $179.42

B $557.38

C $1426.88

D $2140.32

E $2283.01

Use the table for Questions 12 and 13.

Bicycles Advertisement	
Type of Bike	**Regular Price**
12" Bike with Training Wheels	$54.99
16" Trick Bike	$149.99
16" Mountain Bike	$179.99
20" Mountain Bike	$205.99
20" Racing Bike	$267.99

12. If all bikes are 15% off, how much will the 20" mountain bike cost if the sales tax rate is 7%?

A $158.95

B $163.70

C $172.93

D $187.35

E $243.74

13. If all bikes are 20% off, what is the savings between the 16" mountain bike and the 20" mountain bike, before tax?

A $20.80

B $26.00

C $44.80

D $49.60

E $62.00

Grade 9

Name _____ Date _____ Class _____

Test Preparation Practice

Number Properties and Operations

12.1.5.b Solve problems involving factors, multiples, or prime factorization.

Solve each problem, and circle the letter of the best answer.

1. Which list shows all the factors of 32?

   A 1, 2, 4, 8, 16, 32

   B 2^5

   C 1, 2, 4, 8, 16

   D 2, 4, 6, 8, 10, 12, 14, 16

   E 2, 4, 8, 16, 32

2. What is the least common multiple of 15 and 40?

   A 5

   B 8

   C 120

   D 240

   E 600

3. When working on a project in math class, the teacher has the class work in groups of 4 or 6 or 8. When no students are absent, there is one student left over after the students have been assigned to groups. Which of the following could be the number of students in the math class?

   A 18

   B 24

   C 25

   D 33

   E 37

4. The concession stand owner buys hot dogs in packages of 48 and hot dog buns in packages of 32. What is the least number of hot dogs and buns that the owner can buy to have an equal number of each?

   A 64

   B 80

   C 96

   D 128

   E 1536

5. A sporting goods store updates its inventory every time a customer purchases something. Softballs come in packages of 4 or 12. When the inventory sheet showed that the store still had 272 softballs, a sales clerk claimed that this was incorrect. The clerk was correct because 272 is not divisible by what number?

   A 2

   B 4

   C 8

   D 12

   E 16

6. 36 is not a factor of 192 because:

   A 192 is not a prime number.

   B 36 is not a prime number.

   C 6 is not a factor of 192.

   D 9 is not a factor of 192.

   E 4 is not a factor of 192.

Grade 9

7. A bicycle manufacturing company inspects its bicycles as they are made on the assembly line. The table below shows the inspector, what bicycle is checked, and what part of the bicycle is inspected.

Inspector	Bicycle Number	Inspection
Judith	every 10th	paint
Mitchell	every 25th	brakes

Some bicycles could receive both inspections. If 2000 bicycles are produced in a day, how many of them receive both inspections?

A 10

B 25

C 40

D 50

E 80

8. Mrs. Nicholson's class has 24 students and Mr. Salkil's class has 30 students. Each class is having a competition so each class is divided into several teams. What is the greatest number of people on a team if both classes have teams that are equal in size?

A 2 people

B 3 people

C 6 people

D 9 people

E 12 people

9. Margot walks her neighbor's dog every 3rd day. Her friend Karen walks her neighbor's dog every 5th day. They both walk the dogs on July 1st. On what date will they next both walk the dogs?

A July 3rd

B July 5th

C July 15th

D July 16th

E August 1st

10. Which is the prime factorization of 404?

A 404 **D** $50 \cdot 2 + 4$

B $4 \cdot 0 \cdot 4$ **E** $101 \cdot 2^2$

C $2 \cdot 2 \cdot 100$

11. You have a collection of 36 model trains and 21 model cars. You want to arrange your collection on shelves with the same number of models on each shelf. What is the smallest number of shelves you will need?

A 3 shelves

B 4 shelves

C 9 shelves

D 16 shelves

E 36 shelves

12. Tyson has 30 red carnations and 48 white carnations. Tyson needs to create some flower arrangements with the same number of red and white carnations. What is the greatest number of arrangements he can make?

A 2 **D** 8

B 3 **E** 12

C 6

13. A store is having a promotion where random customers who enter the store will receive a gift. The table below shows the gifts and how often they are won. Which customer will be the first to win both a free hat and gift card?

Gift	Customer
Free Hat	Every 12th customer
Free $10 gift card	Every 25th customer

A 20th customer

B 37th customer

C 60th customer

D 200th customer

E 300th customer

Grade 9

Test Preparation Practice
Measurement

12.2.1.c Estimate or compare perimeters or areas of two-dimensional geometric figures.

Solve each problem, and circle the letter of the best answer.

1. What is the approximate circumference of a DVD?

   A 6 inches

   B 9 inches

   C 13 inches

   D 18 inches

   E 24 inches

2. Manuel is renting a storage unit for his classic car. Approximately what area of the floor will the car cover?

   A 40 feet

   B 40 square feet

   C 15 square feet

   D 75 feet

   E 75 square feet

3. Felix is putting up a fence. The back yard measures 24 ft by 32 ft. The fence will be set 4 feet in from the edge on all four sides. What is the approximate perimeter of the enclosed yard?

   A 1000 feet

   B 750 feet

   C 250 feet

   D 100 feet

   E 75 feet

Use the following information for Questions 4–7.

The radius of a circle, the height of a triangle, and the width of a rectangle are all the same length. The base of the triangle and the length of the rectangle are both twice the radius of the circle.

4. Which lists the figures in order of increasing area?

   A circle, triangle, rectangle

   B triangle, rectangle, circle

   C rectangle, circle, triangle

   D triangle, circle, rectangle

   E circle, rectangle, triangle

5. Which lists the figures in order of increasing perimeter or circumference?

   A circle, triangle, rectangle

   B triangle, rectangle, circle

   C rectangle, circle, triangle

   D triangle, circle, rectangle

   E circle, rectangle, triangle

6. If the base of the tiangle is 6 inches, what is the approximate perimeter of the triangle?

   A 8 inches

   B 10 inches

   C 12 inches

   D 14 inches

   E 18 inches

7. If the diameter of the circle is 8 feet, what is the approximate area of the circle?

   A 5 square feet

   B 16 square feet

   C 24 square feet

   D 32 square feet

   E 48 square feet

Grade 9

8. The distance from home plate to the pitcher's mound is about 20 yards. About how many square yards is a baseball diamond?

   A 9 yd^2

   B 90 yd^2

   C 900 yd^2

   D 4200 yd^2

   E 9000 yd^2

9. What is the approximate area of the bottom of a box in which you are to pack 12 soup cans in 4 rows of 3 cans each?

   A 50 square inches

   B 100 square inches

   C 150 square inches

   D 200 square inches

   E 250 square inches

10. A rectangular design has a single stripe from the upper left corner to the lower right corner. The length of the stripe is about 5.2 cm and the length of the rectangle is about 4 cm. What is the approximate perimeter of the rectangular design?

    A 18 cm

    B 14 cm

    C 12 cm

    D 9 cm

    E 6 cm

Use the following information for Questions 11–13.

A pizza shop displays the following chart of pizza diameters in its store window.

Pizza Sizes	
Small	8 in.
Medium	10 in.
Large	12 in.
X-Large	14 in.

11. What is the approximate circumference of a small pizza pan?

    A 12 inches

    B 18 inches

    C 16 inches

    D 24 inches

    E 48 inches

12. What is the approximate difference in area of a medium and an X-large pizza?

    A 72 square inches

    B 24 square inches

    C 48 square inches

    D 12 square inches

    E 96 square inches

13. What is the approximate difference in circumference of a medium and a large pizza?

    A 2 inches

    B 4 inches

    C 6 inches

    D 8 inches

    E 10 inches

Grade 9

Name _____ Date _____ Class _____

Test Preparation Practice
Measurement

12.2.1.d Estimate or compare volume or surface area of three-dimenional figures.

Solve each problem, and circle the letter of the best answer.

1. How much greater is the volume of a waffle cone with height 18 cm and radius 5 cm than a waffle cone with a height of 9 cm and a radius of 5 cm?

   A 235.5 cm^3

   B 353.25 cm^3

   C 471 cm^3

   D 706.5 cm^3

   E 1413 cm^3

2. A number cube has side lengths of 2 cm. A decorative cube has side lengths of 6 cm. How much greater is the surface area of the decorative cube than the surface area of the number cube?

   A 9 times greater

   B 16 times greater

   C 24 times greater

   D 192 times greater

   E 216 times reater

3. About how many cubic inches of yogurt can fit in a cylindrical container with a radius of 2 inches and a height of 5 inches?

   A 20π in^3

   B 18π in^3

   C 15π in^3

   D 10π in^3

   E 8π in^3

4. Grant made a paper-mache sphere with a 2-ft diameter. Ashley made her sphere with a 16-inch radius. Whose sphere has the greater volume, and by about how many times is it greater?

   $\left(\text{Hint: } V = \dfrac{4\pi r^3}{3}\right)$

    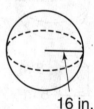

   16 in.

   A Grant; almost 0.5 times as big

   B Grant; at least 2 times as big

   C Ashley; about 1.5 times bigger

   D Ashley; exactly 2 times bigger

   E Ashley; almost 2.5 times bigger

5. A label needs to be wrapped around the cylindrical part of the medicine bottle shown. What is the approximate surface area of this part of the bottle?

   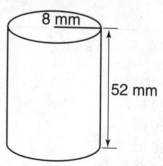

   A 1000 mm^2

   B 1500 mm^2

   C 2000 mm^2

   D 2500 mm^2

   E 3000 mm^2

Grade 9

Name _____ Date _____ Class _____

6. Which figure has the greatest volume?

A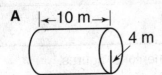

B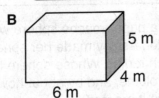
5 m
4 m
6 m

C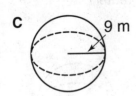
9 m

D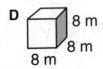
8 m
8 m
8 m

E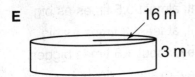
16 m
3 m

7. If Paul wraps a gift in a box that has each dimension 3 times the dimensions of the box shown, what amount of wrapping paper will he need?

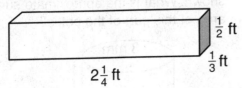
$\frac{1}{2}$ ft
$\frac{1}{3}$ ft
$2\frac{1}{4}$ ft

A 1 ft²
B 10 ft²
C 30 ft²
D 36 ft²
E 42 ft²

8. A rectangular prism has a height of 8 m, a width of 5 m and a length of 12 m. If the length and width of the prism are doubled, how does this affect the volume of the original prism?

A The volume of the new prism is the same as the original prism.

B The volume of the new prism is 0.5 times as large as the original prism.

C The volume of the new prism is 2 times as large as the original prism.

D The volume of the new prism is 4 times as large as the original prism.

E The volume of the new prism is 8 times as large as the original prism.

9. Estimate the volume of the triangular prism.

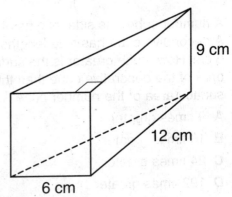
9 cm
12 cm
6 cm

A 72 cm³
B 220 cm³
C 360 cm³
D 420 cm³
E 650 cm³

10. The surface area of a sphere with a diameter of 8 inches is about 201 square inches. About how many times greater is the surface area of a sphere with a diameter of 32 inches?

A 9 times greater
B 16 times greater
C 32 times greater
D 201 times greater
E 3200 times greater

Grade 9

Test Preparation Practice
Measurement

> **12.2.1.e** Solve problems involving the coordinate plane such as the distance between two points, the midpoint of a segment, or slopes of perpendicular or parallel lines.

Solve each problem, and circle the letter of the best answer.

1. Which line on the graph has a negative slope?

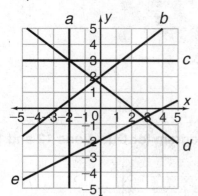

   A line a **D** line d

   B line b **E** line e

   C line c

2. The line shown contains the points $(-2, 4)$ and $(2, -2)$. What is the slope of the line?

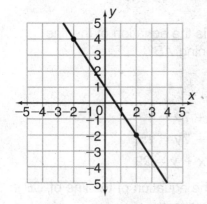

   A $-\dfrac{3}{2}$ **D** 1

   B $-\dfrac{2}{3}$ **E** $\dfrac{3}{2}$

   C $\dfrac{2}{3}$

3. What is the equation of the line?

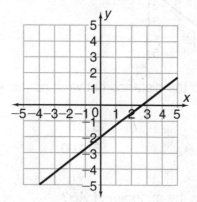

   A $y = 3x - 4$

   B $y = 3x + 4$

   C $y = x - 2$

   D. $y = \dfrac{3}{4}x + 2$

   E $y = \dfrac{3}{4}x - 2$

4. What is the slope of the line perpendicular to the line $y = \dfrac{1}{2}x + 6$?

   A -2

   B $-\dfrac{1}{2}$

   C $\dfrac{1}{6}$

   D $\dfrac{1}{2}$

   E $2\dfrac{1}{6}$

5. What is the equation of the line that is parallel to $y = 4x + 1$ and passes through the point $(-3, 4)$?

   A $y = 4x - 8$

   B $y = 4x - 7$

   C $y = 4x + 1$

   D $y = 4x + 16$

   E $y = 4x - 20$

Grade 9

Use the graph for Questions 6 and 7.

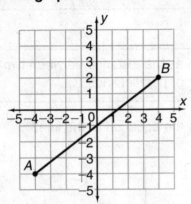

6. What is the distance between points A and B?

A 2   D 10

B 4   E 100

C 8

7. What is the midpoint of $\overline{AB}$?

A $(0, -1)$

B $(-1, 0)$

C $(-2, 2)$

D $(0, -2)$

E $(0, 1)$

8. What is the distance between points $(4, 6)$ and $(7, -6)$?

A 8

B 12

C $3\sqrt{17}$

D $\sqrt{157}$

E $5\sqrt{31}$

9. Find the value of t so that the slope of the line joining points $(t, -4)$ and $(5, 3)$ is $\frac{1}{2}$.

A $t = -9$

B $t = 7$

C $t = 8$

D $t = 9$

E $t = 19$

10. The vertices of a right triangle are $(-3, -1)$, $(-3, 3)$ and $(1, 3)$. What is the length of the hypotenuse?

A 16   D $2\sqrt{5}$

B $4\sqrt{2}$ E 2

C 4

11. What is the slope of a line that contains the points $(9, 3)$ and $(4, 3)$?

A undefined

B $-\frac{5}{6}$

C 0

D $\frac{6}{13}$

E $\frac{6}{5}$

Use the diagram for Questions 12 and 13.

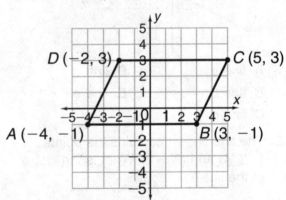

12. What is the equation of the line containing $\overline{AB}$?

A $y = -1$

B $x = -1$

C $2x + 7y = 1$

D $2x + 7y = 15$

E $-7x + y = 29$

13. Find the equation of the line of $\overline{CB}$ in standard form.

A $y = -1$

B $x = -1$

C $y = 2x - 7$

D $y = 2x - 13$

E $y = \frac{1}{2}x - 4$

Name _____ Date _____ Class _____

Test Preparation Practice
Measurement

12.2.1.h Solve mathematical or real-world problems involving perimeter or area of plane figures such as polygons, circles, or composite figures.

Solve each problem, and circle the letter of the best answer.

Use the figure shown for Questions 1 and 2.

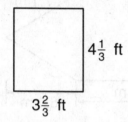

$4\frac{1}{3}$ ft

$3\frac{2}{3}$ ft

1. Determine the perimeter of the rectangle.

   A $6\frac{1}{3}$ ft

   B 8 ft

   C $12\frac{1}{3}$ ft

   D 16 ft

   E $18\frac{2}{3}$ ft

2. Determine the area of the rectangle.

   A $6\frac{1}{5}$ ft^2

   B 13 ft^2

   C $13\frac{7}{9}$ ft^2

   D $15\frac{8}{9}$ ft^2

   E $17\frac{1}{3}$ ft^2

Use the figure shown for Questions 3–5.

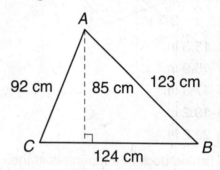

3. Determine the perimeter of triangle *ABC*.

   A 247 cm

   B 254 cm

   C 339 cm

   D 399 cm

   E 424 cm

4. Determine the area of triangle *ABC*.

   A 104.5 cm^2

   B 209 cm^2

   C 5270 cm^2

   D 10,540 cm^2

   E 15,810 cm^2

5. A rectangle has a length of *CB* and its width is the height of triangle *ABC*, what is the area of the rectangle?

   A 104.5 cm^2

   B 209 cm^2

   C 5270 cm^2

   D 10,540 cm^2

   E 15,810 cm^2

Grade 9

Name _____ Date _____ Class _____

6. A pond is represented on a bluerint as shown. Find the perimeter of the pond.

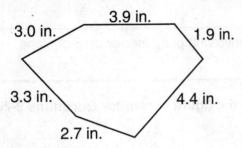

3.0 in.
3.9 in.
1.9 in.
3.3 in.
4.4 in.
2.7 in.

A 15.3 in.

B 15.9 in.

C 17.8 in.

D 19.2 in.

E 22.2 in.

7. A fishing dock at a marina is in the shape of a trapezoid as shown. Determine the area of the dock.

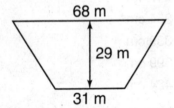

68 m
29 m
31 m

A 49.5 m^2

B 1178 m^2

C 1435.5 m^2

D 2356 m^2

E 2871 m^2

8. Determine the area of the circle. (Use $\pi = 3.14$.)

5 in.

A 39.25 in.^{2}

B 78.5 in.^{2}

C 157 in.^{2}

D 314 in.^{2}

E 628 in.^{2}

9. Determine the area of a rectangle that is 4.5 cm long by 0.8 cm wide.

A 3.6 cm^2

B 4.8 cm^2

C 5.3 cm^2

D 7.2 cm^2

E 10.6 cm^2

Use the figure for Questions 10 and 11.

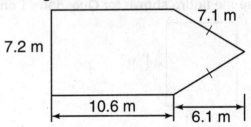

7.1 m
7.2 m
10.6 m
6.1 m

10. Determine the perimeter of the figure.

A 24.9 m

B 41.6 m

C 42.6 m

D 48.6 m

E 54.8 m

11. Determine the area of the figure.

A 44.26 m^2

B 98.28 m^2

C 116.39 m^2

D 232.78 m^2

E 465.55 m^2

Grade 9

Name _____ Date _____ Class _____

Test Preparation Practice
Measurement

12.2.1.j Solve problems involving volume or surface area of rectangular solids, cylinders, cones, pyramids, prisms, spheres, or composite figures.

Solve each problem, and circle the letter of the best answer.

1. Determine the volume of a waffle cone with height 18 cm and radius of 5 cm. Round to the nearest cubic centimeter.

   A 94 cm^3

   B 188 cm^3

   C 282 cm^3

   D 471 cm^3

   E 1413 cm^3

Use the picture for Questions 2 and 3.

2 in.

2. Determine the surface area of one fair die.

   A 4 in.^{2}

   B 6 in.^{2}

   C 8 in.^{2}

   D 16 in.^{2}

   E 24 in.^{2}

3. Determine the volume of both fair dice.

   A 4 in.^{3}

   B 6 in.^{3}

   C 8 in.^{3}

   D 16 in.^{3}

   E 24 in.^{3}

4. Determine the volume of a sphere with a radius of 2 ft. $\left(\text{Hint: } V = \dfrac{4\pi r^3}{3}\right)$.

   A 16.75 ft^3

   B 33.49 ft^3

   C 50.24 ft^3

   D 110.48 ft^3

   E 301.44 ft^3

Use the figure shown for Questions 5 and 6.

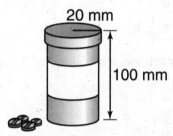

20 mm

100 mm

5. Determine the surface area of the medicine bottle given a height of 100 mm and a radius of 20 mm.

   A 2512 mm^2

   B 6908 mm^2

   C 10,048 mm^2

   D 12,560 mm^2

   E 15,072 mm^2

6. Determine the volume of the medicine bottle.

   A 6280 mm^3

   B 1256 mm^3

   C 31,400 mm^3

   D 125,600 mm^3

   E 131,880 mm^3

Grade 9

Name _____ Date _____ Class _____

Use the figure shown for Questions 7 and 8.

$2\frac{1}{4}$ ft

$\frac{1}{3}$ ft $\frac{1}{2}$ ft

7. Determine the surface area of the gift.

   A $2\frac{1}{12}$ ft^2

   B $2\frac{1}{6}$ ft^2

   C $2\frac{11}{12}$ ft^2

   D 3 ft^2

   E $4\frac{1}{12}$ ft^2

8. Determine the volume of the gift.

   A $\frac{1}{4}$ ft^3

   B $\frac{3}{8}$ ft^3

   C $\frac{5}{8}$ ft^3

   D $1\frac{1}{8}$ ft^3

   E $3\frac{3}{8}$ ft^3

9. The volume V of a sphere is given by the formula $V = \frac{4}{3}\pi r^3$. If the radius r of the sphere is equal to $\frac{1}{2}s$, what is the volume of the sphere?

   A $\frac{2}{3}\pi s^3$

   B $2\pi s$

   C πs^3

   D $\frac{1}{3}\pi s$

   E $\frac{1}{6}\pi s^3$

10. Determine the volume of the triangular prism.

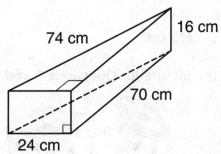

74 cm   16 cm

70 cm

24 cm

   A 192 cm^3
   B 840 cm^3
   C 13,440 cm^3
   D 26,880 cm^3
   E 994,560 cm^3

11. The volume of a hemisphere is 718.38 m^3. What is the radius?

   A 2 m
   B 7 m
   C 12 m
   D 14 m
   E 21 m

Grade 9

Test Preparation Practice
Measurement

12.2.1.l Solve problems involving rates such as speed, density, population density, or flow rates.

Solve each problem, and circle the letter of the best answer.

1. The state of Michigan covers an area of 96,810 square miles. In 2004, the population of Michigan, according to the United States Census Bureau, was 9,938,444 people. What is the population density of Michigan per square mile?

   A 9 people per square mile

   B 10 people per square mile

   C 103 people per square mile

   D 130 people per square mile

   E 309 people per square mile

2. A boat travels 195 miles for 13 hours. What is the average rate of speed of the boat?

   A 208 mph

   B 182 mph

   C 50 mph

   D 15 mph

   E 8 mph

3. The total area of Rhode Island is 1545 square miles with a current population of about 1,050,000. What is the approximate population density of Rhode Island?

   A 1 person per square mile

   B 54 people per square mile

   C 525 people per square mile

   D 680 people per square mile

   E 2225 people per square mile

Use the information below for Questions 4 and 5.

Density is defined as the amount of matter present in a given volume of a substance, or mass divided by volume.

4. A student finds that 4.53 kilograms of a substance has a volume of 225 milliliters. What is the density of the substance?

   A 0.05 kg/L

   B 5.00 kg/L

   C 20.13 kg/L

   D 47.22 kg/L

   E 50.00 kg/L

5. At 20°C, the density of pure silver is 10.5 g/cm^3. If a sample of silver has a volume of 125 cm^3, what is the mass of the silver sample?

   A 0.05 g

   B 11.9 g

   C 625.9 g

   D 1312.5 g

   E 1875.5 g

6. At Frost High School, there are 675 students, 42 teachers, and 5 custodians. What is the student to teacher ratio at this school?

   A 8 : 1

   B 16 : 1

   C 18 : 1

   D 32 : 1

   E 42 : 1

Grade 9

Name _____ Date _____ Class _____

7. Rebecca has a 250-gram sample of each substance listed in the table. Which substance has the greatest volume?

Substance	Physical State	Density (g/cm^3)
silver	solid	10.5
magnesium	solid	1.74
iron	solid	7.87
aluminum	solid	2.70
copper	solid	8.96

  A Silver

  B Magnesium

  C Iron

  D Aluminum

  E Copper

8. A bicyclist rides at an average rate of 10 miles per hour. How does it take the bicyclist to travel 78 miles?

  A 7.8 h

  B 78 h

  C 450 h

  D 468 h

  E 780 h

Use the information below for Questions 9 and 10.

A mountain climber hikes 1350 feet in one hour. He then rests for 45 minutes before hiking an additional 1650 feet in 100 minutes.

9. What is the climber's average rate of speed before his 45 minute break?

  A 6.5 ft/min

  B 13.5 ft/min

  C 22.5 ft/min

  D 675 ft/min

  E 1350 ft/min

10. What is the climber's overall average rate for his entire hike?

  A 22.5 ft/min

  B 18.8 ft/min

  C 14.6 ft/min

  D 9.3 ft/min

  E 6 ft/min

Use the information below for Questions 11 and 12.

A pump and piping system is used to empty an oil tank at a rate of 125 gallons per minute. The time t needed to empty the tank is represented by the equation $t = \dfrac{c}{125}$, where c is the capacity of the tank.

11. How long will it take to empty a 12,000 gallon oil tank?

  A $\dfrac{3}{5}$ hour

  B $\dfrac{4}{5}$ hour

  C 1 hour

  D 1.25 hours

  E $1\dfrac{3}{5}$ hours

12. Suppose the pump is not working properly and the truck driver only has 2.5 hours to empty the 12,000 gallon tank. What is the slowest the pump can function to completely empty the tank in the given amount of time?

  A 45 gallons per minute

  B 80 gallons per minute

  C 125 gallons per minute

  D 200 gallons per minute

  E 4800 gallons per minute

Grade 9

Test Preparation Practice

Measurement

> **12.2.1.m** Use trigonometric relations in right triangles to solve problems.

Solve each problem, and circle the letter of the best answer.

Figures are not drawn to scale.

Use the right triangle for Questions 1 to 3.

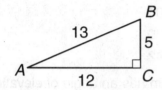

1. What is the sine of $\angle A$?

  A $\frac{5}{13}$

  B $\frac{5}{12}$

  C $\frac{12}{13}$

  D $\frac{13}{12}$

  E $\frac{12}{5}$

2. What is the cosine of $\angle A$?

  A $\frac{5}{13}$

  B $\frac{5}{12}$

  C $\frac{12}{13}$

  D $\frac{13}{12}$

  E $\frac{12}{5}$

3. What is the tangent of $\angle A$?

  A $\frac{5}{13}$

  B $\frac{5}{12}$

  C $\frac{12}{13}$

  D $\frac{13}{12}$

  E $\frac{12}{5}$

4. What is the length of the hypotenuse of the right triangle? Round your answer to the nearest hundredth.

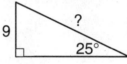

  A 3.80

  B 8.16

  C 9.93

  D 19.3

  E 21.30

5. The length of side a is 24. The tangent of $\angle A$ is $\frac{12}{5}$. What is the length of the hypotenuse?

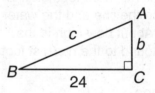

  A 5

  B 12

  C 26

  D $\sqrt{313}$

  E $12\sqrt{5}$

Grade 9

Name _____ Date _____ Class _____

6. Which of the following has the same value as sin A?

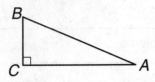

- **A** sin B
- **B** cos A
- **C** cos B
- **D** tan A
- **E** tan B

7. What is the length of x? Round your answer to the nearest hundredth.

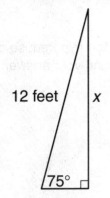

12 feet   x

75°

- **A** 3.11 feet
- **B** 9.32 feet
- **C** 10.42 feet
- **D** 11.59 feet
- **E** 12.42 feet

8. A boat is pulling a parasailer. The line to the parasailer is 1200 feet long. The angle between the line and the water is about 32°. About how high is the parasailer? Round to the nearest foot.

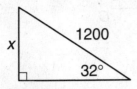

x   1200

32°

- **A** 640 feet
- **B** 749 feet
- **C** 1017 feet
- **D** 1415 feet
- **E** 2264 feet

9. What is the length of x? Round your answer to the nearest hundredth.

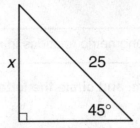

x   25

45°

- **A** 17.00
- **B** 17.68
- **C** 24.04
- **D** 25.00
- **E** 37.42

10. The sun has an angle of elevation of 42°. How long a shadow, x, does a person 64 inches tall make?

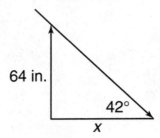

64 in.

42°

x

- **A** 42.82 in.
- **B** 47.56 in.
- **C** 57.63 in.
- **D** 71.08 in.
- **E** 86.12 in.

11. An 18-inch diagonal of a rectangular picture frame creates a 30°–60°–90° triangle. What is the length of the longest side of the picture frame?

- **A** 9.0 in.
- **B** 15.59 in.
- **C** 20.78 in.
- **D** 31.18 in.
- **E** 36.0 in.

Grade 9

Name _____ Date _____ Class _____

Test Preparation Practice
Measurement

12.2.2.a Select or use appropriate type of unit for the attribute being measured such as volume or surface area.

Solve each problem, and circle the letter of the best answer.

Use the information below for Questions 1 and 2.

A traffic cone has a perpendicular height of 28 inches. The circumference of the opening in the base of the cone is 24.7 inches.

1. To determine the volume of a traffic cone, its radius must be measured. What unit corresponds to the radius of a traffic cone?

   A cubic feet
   B square centimeter
   C inch
   D gram
   E quart

2. A student determines that the volume of the cone is 452.8 in². Why is this NOT correct?

   A The unit should be inches.
   B The volume is 522 in².
   C The measure for diameter was used.
   D The unit should be cubic inches.
   E Cannot be determined

3. The volume of a hemisphere is 661.52 ft³. What unit of measure is NOT acceptable for the radius?

   A feet
   B inch
   C yard
   D meter
   E gram

4. Which units could be used to express the surface area of a tent?

   A square meters
   B cubic yards
   C inches
   D cubic inches
   E centimeters

5. A rectangular gift box has a volume of 3625 cubic inches. What is the most likely unit of measure of each side?

   A square inches
   B cubic feet
   C square feet
   D inches
   E kilograms

6. What is the appropriate unit of measure for the amount of wrapping paper needed to wrap a gift box?

   A yards
   B cubic feet
   C cubic inches
   D square inches
   E rolls

Grade 9

7. The side lengths of an irregularly-shaped marble slab is shown. In order to find the volume of the marble slab, what dimension do you need to know?

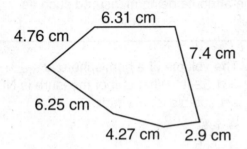

6.31 cm

4.76 cm

7.4 cm

6.25 cm

4.27 cm   2.9 cm

A Radius

B Capacity

C Surface Area

D Perimeter

E Height

8. Suppose that you need to know the amount of material needed to make a basketball. You already know that the distance around the basketball is 69 centimeters. What other information do you need?

A The volume in cubic centimeters

B The radius in centimeters

C The surface area in square centimeters

D The weight in grams

E The circumference in centimeters

9. The volume of Earth can be expressed using what unit of measure?

A square miles

B kilograms

C micrometers

D cubic meters

E light-years

10. When determining the difference in volumes of the two cans shown, what unit of measure, if any, should correspond to your answer?

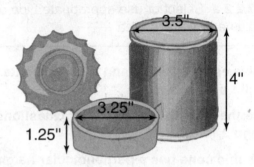

3.5"

4"

3.25"

1.25"

A inches

B square inches

C cubic inches

D fluid ounces

E No unit of measure

11. Determine the surface area of a roll of paper towels with a height of 35.4 cm and a diameter of 20 cm. (Hint: Use $SA = 2\pi r^2 + 2\pi rh$ and $\pi = 3.14$.)

20 cm

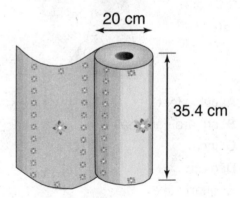

35.4 cm

A 2851.12 in^2

B 2851.12 m^2

C 2851.12 lb^2

D 2851.12 kg^2

E 2851.12 cm^2

Grade 9

Name _____ Date _____ Class _____

Test Preparation Practice
Measurement

12.2.2.b Solve problems involving conversions within or between measurement systems, given the relationship between the units.

Solve each problem, and circle the letter of the best answer.

1. Justin is 5 ft 9 in. tall and Caleb is 71 in. tall. Which of the following statements is true?

   A Justin is 5 inches taller than Caleb.

   B Caleb is 3 inches taller than Justin

   C The boys are the same height.

   D Caleb is 3 inches shorter than Justin.

   E Justin is 2 inches shorter than Caleb.

2. A home filtered water system is supplied by 5 gallon jugs. How many 8 fluid ounce glasses of water can be filled from the jug?

   A 640 glasses

   B 80 glasses

   C 100 glasses

   D 140 glasses

   E 640 glasses

Use the recipe shown below for Questions 3 and 4.

Garden Vegetable Juice

2 quarts tomato juice $\frac{1}{4}$ cup lemon juice

$\frac{3}{4}$ cup carrot juice 32 oz cold water

3. What is the smallest container that will hold the ingredients?

   A 3 quart

   B 5 quart

   C 6 quart

   D 1 gallon

   E 2 gallon

4. If the recipe is to be tripled, how many ounces of carrot juice are needed?

   A 12

   B 16

   C 18

   D 20

   E 24

5. What is the difference in length between 76 cm and 1 meter?

   A 14 cm

   B 24 cm

   C 66 cm

   D 86 cm

   E 924 cm

6. In an 8-hour shift, Pam actually worked 530 minutes. How much overtime did she work?

   A 20 minutes

   B $\frac{1}{2}$ hour

   C 40 minutes

   D 50 minutes

   E 1 hour

7. A crafter buys 3 meters of ribbon. She uses 40 centimeters for one project and 70 centimeters for another project. How many meters of ribbon does she have left?

   A 1900 m

   B 190 m

   C 19 m

   D 1.9 m

   E 0.19 m

Grade 9

8. A nurse must record a newborn baby's weight in ounces to calculate the proper amount of medication. If the newborn weighs $6\frac{3}{4}$ pounds, how many ounces does the newborn weigh?

A 96 ounces

B $96\frac{3}{4}$ ounces

C 108 ounces

D 120 ounces

E $216\frac{1}{4}$ ounces

9. The record for a javelin throw is 247 ft 10 in. for women and 342 ft 2 in. for men. How much further is the men's record?

A 95 ft 8 in.

B 95 ft 4 in.

C 94 ft 4 in.

D 94 ft 2 in.

E 93 ft 6 in.

10. Manuel takes 500 milligrams of vitamin C daily. How many grams does he take in a week?

A 3.5 g

B 35 g

C 350 g

D 3500 g

E 3,500,000 g

11. The rate of one knot equals one nautical mile per hour. One nautical mile is 1852 meters. What is the speed in meters per second of a ship traveling at 25 knots?

A 46.3 m/s

B 74.08 m/s

C 12.86 m/s

D 128.6 m/s

E 4.63 m/s

12. Tonya plants 4 seedlings of pine trees that grow about 18 inches in a year, for a windbreak. About how many feet will each seedling be in 5 years?

A $6\frac{1}{2}$ ft

B $7\frac{1}{2}$ ft

C 9 ft

D 16 ft

E 30 ft

13. Natalie baby-sits her neighbor's children each afternoon. How many hours did she work for the week shown?

Natalie's Baby-sitting Records	
Weekday	**Time**
Monday	1 hour 35 minutes
Tuesday	2 hours 20 minutes
Wednesday	$1\frac{1}{4}$ hours
Thursday	55 minutes
Friday	$1\frac{2}{3}$ hours

A 6 hours 50 minutes

B 7 hours

C $7\frac{1}{2}$ hours

D $7\frac{3}{4}$ hours

E $8\frac{1}{3}$ hours

14. A cylindrical tank has a volume of 640 ft^3. What is this volume in cubic inches?

A 53.33 in^3

B 7680 in^3

C 23,040 in^3

D 96,160 in^3

E 1,105,920 in^3

Grade 9

State Test Preparation Practice
Measurement

12.2.2.f Construct or solve problems (e.g. number of rolls needed for insulating a house) involving scale drawings.

Solve each problem, and circle the letter of the best answer.

Figures are not drawn to scale.

1. A model dinosaur with a scale of 1.5 in. = 10 feet is needed for an animated film. If the model dinosaur is 4.5 inches tall, what is the actual size of the dinosaur?

   A 15 ft

   B 25 ft

   C 30 ft

   D 45 ft

   E 60 ft

2. The model of the front face of a roof is shown.

   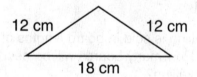

   Scale 2 cm: 5 m

   If one roof shingle covers 0.5 m², about how many shingles do you need to cover the front face of the actual roof?

   A 432

   B 447

   C 755

   D 893

   E 1510

3. A pair of binoculars magnifies an object by 5 times its actual size. For example, the binoculars make an object 100 feet away appear to be 20 feet away. When viewing a concert using the binoculars, the drummer appears to be 15 feet away. What is the viewer's actual distance from the drummer?

   A 3 ft

   B 5 ft

   C 30 ft

   D 75 ft

   E 100 ft

4. You have a 4-inch by 6-inch photo that you want to enlarge proportionally and set on premium poster paper. The premium poster paper costs $0.03 per square inch. If you have $20 to spend on the paper, what is the greatest size of poster you can buy?

   A 12 in. by 18 in.

   B 16 in. by 24 in.

   C 20 in. by 30 in.

   D 24 in. by 36 in.

   E 28 in. by 42 in.

Grade 9

Use the diagram and information for Questions 5–7.

The floor plan of a garage is shown. The garage height is 16 ft.

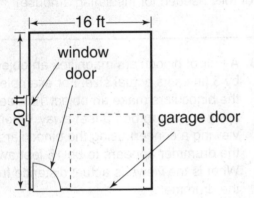

5. The 14-in. by 14-in. square tile pieces for the garage floor cost $3.50 each. About how much money will it cost to cover the garage floor with these tiles?

A $320   D $1120

B $435   E $4480

C $826

6. Three walls (not including the wall with the garage door) will be filled with insulation. Each roll of insulation is 3 feet wide by 18 feet long. About how many rolls of insulation are needed to insulate the three walls?

A 56   D 19

B 27   E 12

C 25

7. The window is 3 feet by 5 feet, the door is 3 feet by 9 feet, and the garage door is 8 feet by 13 feet. If one can of paint covers 400 square feet, about how many cans of paint will be needed to cover the interior walls with two coats of paint?

A 1   D 10

B 2   E 12

C 5

Use the diagram and information for Questions 8 and 9.

The floor plan of a deck is shown. The scale of the diagram is 1 in. = 10 ft. The measures of the scale drawing are given.

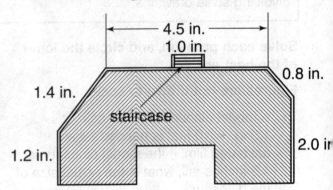

8. The deck will have a railing around the outer edge, except for the staircase opening. About how much railing will be needed?

A 9 ft

B 19 ft

C 45 ft

D 89 ft

E 99 ft

9. The railing is sold in board lengths of 3 yards. How many boards will be needed for the railing?

A 10

B 18

C 22

D 27

E 30

Grade 9

Test Preparation Practice
Geometry

12.3.2.c Perform or describe the effect of a single transformation on two- and three-dimensional geometric shapes (reflections across lines of symmetry, rotations, translations, and dilations).

Solve each problem, and circle the letter of the best answer.

1. Which of the following terms is not a term representing a transformation?

   A reflection

   B dilation

   C translation

   D rotation

   E transportation

2. How many lines of symmetry does the figure have?

   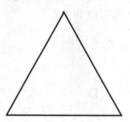

   A 0   **D** 3

   B 1   **E** 4

   C 2

3. What is the transformation from Figure 1 to Figure 2?

   Figure 1   Figure 2

   A translation   **D** symmetry

   B reflection   **E** dilation

   C rotation

4. This figure appears to be symmetric with respect to:

   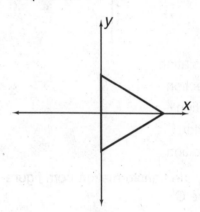

   A the x-axis only

   B the y-axis only

   C both the x-axis and y-axis

   D neither the x- nor the y-axis

   E itself

5. Identify the transformation from Figure A to Figure B.

   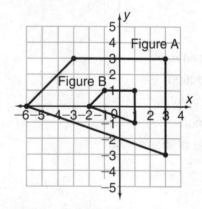

   A magnification

   B translation

   C reflection

   D dilation

   E rotation

Grade 9

Name _____ Date _____ Class _____

6. Identify the transformation from figure A to figure B.

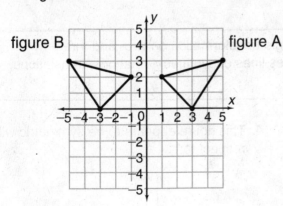

figure B   figure A

A translation

B reflection

C rotation

D dilation

E transition

7. Identify the transformation from figure B to figure C.

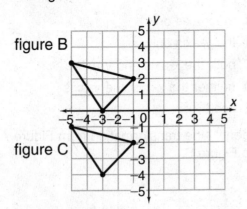

figure B

figure C

A translation

B reflection

C rotation

D dilation

E none of these

8. Identify the transformation from figure A to figure B.

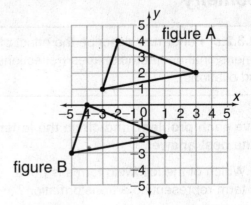

figure A

figure B

A translation

B reflection

C rotation

D dilation

E none of the above

9. Which white trapezoid shows the position of the black trapezoid after being reflected over the line?

A   **D**

B   **E**

C

Grade 9

Test Preparation Practice
Geometry

12.3.2.d Describe the final outcome of successive transformations.

Solve each problem, and circle the letter of the best answer.

1. Identify the transformation from figure A to figure B.

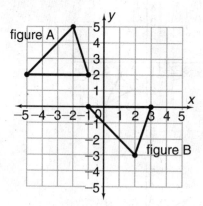

A translation, rotation

B translation, reflection

C reflection, dilation

D reflection, rotation

E dilation, rotation

2. What are the coordinates of the image of A when △ABC is translated 3 units to the left and 2 units down?

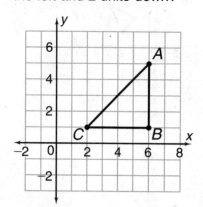

A (3, 7)   **D** (3, 3)

B (9, 3)   **E** (−1, −1)

C (4, 2)

3. What is the image of point (4, −6) when it is reflected across the *x*-axis and translated up 2 units?

A (−6, −2)

B (−4, 8)

C (−4, −4)

D (−6, 6)

E (4, 8)

4. Identify the transformation from triangle *ABC* to triangle *A′B′C′*.

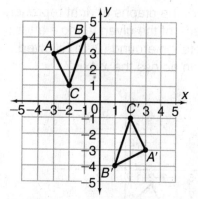

A reflection, reflection

B rotation, reflection

C rotation, translation

D dilation, translation

E translation, translation

Grade 9

Name _____ Date _____ Class _____

5. Point *P* is translated to the right 2 units and down 5 units. What rule describes the translation?

   A $(x, y) \rightarrow (x + 5, y - 2)$

   B $(x, y) \rightarrow (x - 5, y + 2)$

   C $(x, y) \rightarrow (x - 2, y + 5)$

   D $(x, y) \rightarrow (x + 2, y - 5)$

   E $(x, y) \rightarrow (x + 2, y + 5)$

6. Point *A* (2, −3) is translated to the left 3 units and up 4 units. What are the coordinates of *A'*?

   A (−1, 1)

   B (0, 0)

   C (5, −7)

   D (−1, 7)

   E (5, 1)

7. Which of the graphs at right represents the image of the given triangle after a 180° rotation around (0, 0) followed by a reflection across the *x*-axis?

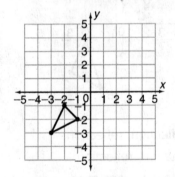

A

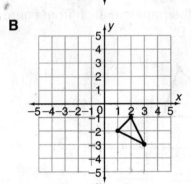

B

C

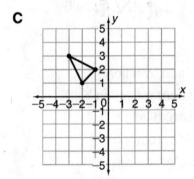

D

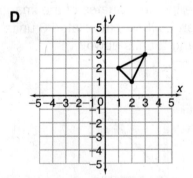

E

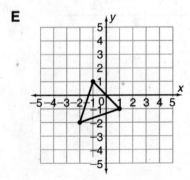

Grade 9

Name _____ Date _____ Class _____

Test Preparation Practice
Geometry

12.3.2.e Justify relationships of congruence and similarity, and apply these relationships using scaling and proportional reasoning.

Solve each problem, and circle the letter of the best answer.

Figures are not drawn to scale.

1. Which best represents a pair of similar figures?

A

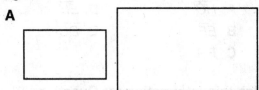

B

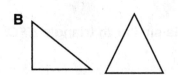

C

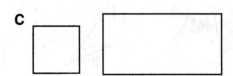

D

E

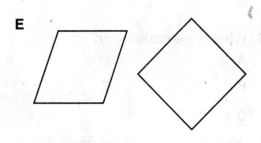

2. Triangle *ABC* is similar to triangle *DEF*. Find the measure of angle *x*.

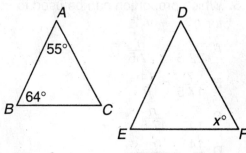

A 32°	**D** 116°
B 61°	**E** 119°
C 64°	

3. The figures below are similar. What is the perimeter of the smaller figure?

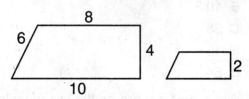

A 12 units	**D** 26 units
B 14 units	**E** 28 units
C 20 units	

4. Which best completes the congruence statement? $\overline{AC} \cong$ _____

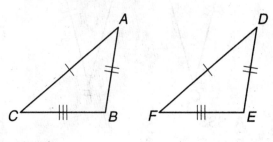

A $\overline{DE}$	**D** $\overline{DF}$
B $\overline{FE}$	**E** $\overline{AB}$
C $\overline{CB}$	

Grade 9

Use the similar figures for Questions 5 and 6.

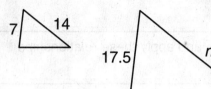

5. Which proportion can be used to solve for *n*?

A $\dfrac{7}{22.5} = \dfrac{n}{17.5}$

B $\dfrac{7}{17.5} = \dfrac{14}{n}$

C $\dfrac{7}{14} = \dfrac{n}{17.5}$

D $\dfrac{14}{7} = \dfrac{n}{22.5}$

E $\dfrac{7}{14} = \dfrac{22.5}{n}$

6. What is the value of *n*?

A 8.75

B 10.9

C 35

D 45

E 54

7. Which theorem proves the two triangles congruent?

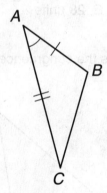
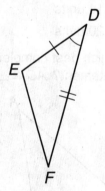

A SSS

B SAS

C AAS

D HL

E ASA

8. Which best completes the congruence statement? $\overline{AB} \cong$ _____

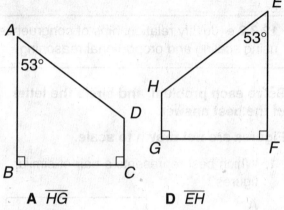

A $\overline{HG}$　　　　　**D** $\overline{EH}$

B $\overline{EF}$　　　　　**E** $\overline{DC}$

C $\overline{FG}$

Use this information for Questions 9 and 10.

Triangle *ABC* is similar to triangle *EFG*.

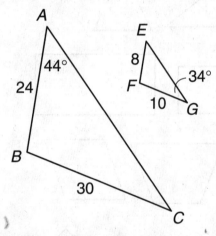

9. What is the scale factor?

A 24　　　　　**D** 3

B 12　　　　　**E** $\dfrac{1}{2}$

C 6

10. What is the measure of angle *B*?

A 34°　　　　　**D** 102°

B 44°　　　　　**E** 132°

C 78°

Grade 9

Test Preparation Practice
Geometry

12.3.3.c Represent problem situations with geometric models to solve mathematical or real-world problems.

Solve each problem, and circle the letter of the best answer.

Figures are not drawn to scale.

1. A circle has a circumference of 50.24 feet. What is the radius of the circle?

   A 6 feet

   B 8 feet

   C $10\frac{1}{2}$ feet

   D 12 feet

   E 16 feet

2. A tile is in the shape of a 30°–60°–90° triangle. The side across from the 30° angle has a measure of 6.25 cm. What is the measure of the hypotenuse of this tile?

   A $\sqrt{3}$ cm

   B 3.125 cm

   C 6.25 cm

   D 12.5 cm

   E 15 cm

3. A school decides to increase the width of its rectangular playground from 25 m to 40 m and the length from 45 m to 60 m. By how much does the perimeter of the playground increase?

   A 30 m   **D** 200 m

   B 40 m   **E** 225 m

   C 60 m

4. If a parallelogram has an area of 240 square inches and the height is 18 inches, what is the length of the base?

   A $6\frac{2}{3}$ inches   **D** 2160 inches

   B $13\frac{1}{3}$ inches   **E** 4320 inches

   C $26\frac{2}{3}$ inches

5. An architect has designed a restaurant balcony as shown by the shaded regions. If the flower boxes are to be placed along the outside edges of each balcony, what length will be covered by the flower boxes, to the nearest tenth?

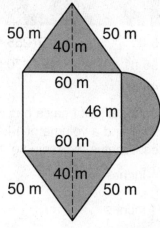

   A 166 m   **D** 272.3 m

   B 236.5 m   **E** 366.6 m

   C 246.0 m

6. A Frisbee has an area of 50.24 square inches. What is the diameter of the Frisbee?

   A 4 inches

   B 8 inches

   C 10 inches

   D 12 inches

   E 16 inches

Grade 9

Name _____ Date _____ Class _____

7. Mrs. DiMoreto purchases the new rug shown below for her living room floor. The rug measures 3 feet by 5 feet. Her living room is 15 feet by 18 feet. How many square feet of living room floor does the rug NOT cover?

A 15 ft^2 D 255 ft^2
B 156 ft^2 E 270 ft^2
C 245 ft^2

8. A cylindrical jar of salsa has a height of 2 inches and a volume of 14.13 cubic inches. What is the radius of the jar?

A $1\frac{1}{2}$ inches

B $2\frac{1}{4}$ inches

C 3 inches

D 5 inches

E 7 inches

9. A shipping company orders a new conveyor belt to move packages. The conveyor belt may be 20 inches longer or shorter than the belt shown below. What is the range of the perimeter, in feet, of the conveyor belt?

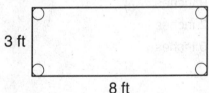

3 ft

8 ft

A $16 \le x \le 56$
B $23 \le x \le 28$
C $76 \le x \le 116$
D $112 \le x \le 152$
E $244 \le x \le 284$

Use the composite figure shown below for Questions 10 and 11.

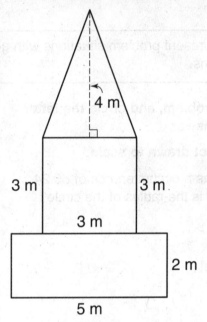

4 m

3 m   3 m

3 m

2 m

5 m

10. What is the perimeter of the composite figure to the nearest tenth?

A 18.5 meters
B 19.3 meters
C 20.5 meters
D 21.3 meters
E 25.5 meters

11. What is the area of the composite figure?

A 8 square meters
B 21 square meters
C 25 square meters
D 42 square meters
E 45 square meters

Grade 9

Test Preparation Practice
Geometry

12.3.3.d Use the Pythagorean theorem to solve problems in two- or three-dimensional situations.

Solve each problem, and circle the letter of the best answer.

Figures are not drawn to scale.

1. In the right triangle shown, find the length of the unknown side.

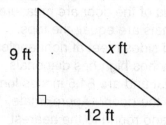

A. $\sqrt{21}$ feet

B 15 feet

C 21 feet

D 54 feet

E 108 feet

2. In the right triangle shown, find the length of the unknown side.

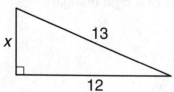

A. 1

B 5

C 25

D 19.69

E 313

3. A light pole is 24 feet high. It is to be stabilized with a guy wire stretched from the top of the pole to a position 7 feet from the base of the pole. How long must the guy wire be to stabilize the pole?

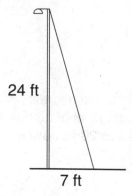

A. 17 feet

B 25 feet

C 27 feet

D 31 feet

E 625 feet

4. A police officer leaves the station and drives 9 miles east, then makes a right turn and drives 21 miles south. How far is the officer from the station to the nearest hundredth?

A 18.45 miles

B 18.46 miles

C 21.93 miles

D 22.85 miles

E 30.01 miles

Grade 9

Name _____ Date _____ Class _____

5. Reba has found that her cell phone will operate if it is no more than 17 miles from a cell phone tower. If Reba lives 8 miles north of the nearest tower, how far west is she able to drive and still operate her cell phone using that tower?

A. 9 miles

B. $2\sqrt{34}$ miles

C. $3\sqrt{14}$ miles

D. 15 miles

E. $\sqrt{353}$ miles

6. What is the length, to the nearest hundredth, of the diagonal of a square whose side measures 20 inches?

A. 24.83 in.

B. 27.12 in.

C. 28.28 in.

D. 29.17 in.

E. 30.00 in.

7. Max's school has a rectangular courtyard that measures 96 meters by 72 meters. How much shorter is the walk from the library to the school along the diagonal sidewalk than through point A?

Library

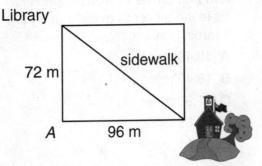

A. 48 m   D 120 m

B 88 m   E 168 m

C 96 m

8. Suppose you are making a sail in the shape of a right triangle. The length of the longest side is 65 ft. The sail is to be 63 ft high. What is the length of the third side of the sail?

A. 2 feet

B. 15 feet

C. 16 feet

D. 90.52 feet

E. 128 feet

9. To ensure that tops, bottoms and sides of a door frame meet at right angles, the diagonals of the door are measured. If the diagonals are equal, the tops, bottoms and sides meet at right angles. A door 78 inches high has diagonals that are equal and are 84.5 inches long. How wide is the door? Approximate the answer and round to the nearest hundredth.

A. 13.5 inches

B. 32.50 inches

C. 55.15 inches

D. 65.91 inches

E. 114.00 inches

10. Which of the following are measures of three sides of a right triangle?

A. 1, 2, 3

B. 4, 7, 8

C. 3, 7, 9

D. 10, 15, 20

E. 9, 12, 15

Grade 9

Test Preparation Practice
Geometry

12.3.4.d Represent two-dimensional figures algebraically using coordinates and/or equations.

Solve each problem, and circle the letter of the best answer.

1. What figure is formed by the intersection of the four lines given below?

$y = 2x + 3$

$y = -\frac{1}{2}x - 1$

$y = 2x - 4$

$y = -\frac{1}{2}x + 4$

A Triangle

B Circle

C Rectangle

D Rhombus

E Trapezoid

2. If one side of a square is contained by the line $y = 4x - 2$, which equation describes a line that could contain one of the other 3 sides?

A $y = \frac{1}{4}x - 2$

B $y = -\frac{1}{4}x + 3$

C $y = -4x - 2$

D $y = 4x - 3$

E $y = \frac{1}{4}x + 2$

3. Which of the following sets of ordered pairs do NOT lie on or within the triangle formed by the intersection of the following lines?

$y = \frac{1}{2}x + 5$

$x = 2$

$y = 1$

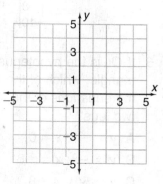

A $\{(0, 2), (-4, 1), (1, 3)\}$

B $\{(-2, 3), (-3, 2), (-4, 1)\}$

C $\{(1, 1), (2, 4), (3, 5)\}$

D $\{(0, 5), (0, 3), (-1, 1)\}$

E $\{(1, 3), (-2, 3), (-3, 1)\}$

4. Which of the following best classifies the type of triangle formed by the intersection of the following lines?

$y = |x + 7|$

$y = 6$

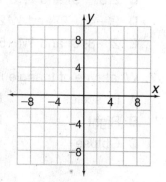

A Isosceles, equiangular

B Right, isosceles

C Obtuse

D Acute, skew

E Obtuse, skew

Grade 9

Use the coordinate grid below for Questions 5–7.

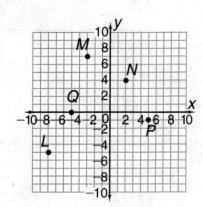

5. If $\overline{QL}$ is the hypotenuse of a right triangle, which of the following could be the equations of the lines containing the two legs?

A $y = -8; x = 0$

B $y = -5; x = -8$

C $y = 5; x = -8$

D $y = x; x = -5$

E $y = 0; x = -8$

6. If $QLZW$ is a regular trapezoid, which of the following indicates where points Z and W are located?

A $Z(-2, -5)$ and $W(-2, 0)$

B $Z(3, -5)$ and $W(0, 0)$

C $Z(8, -5)$ and $W(6, 0)$

D $Z(-4, -5)$ and $W(2, 2)$

E $Z(-8, 6)$ and $W(-5, 2)$

7. What type of triangle is QMP?

A Isosceles   **D** Acute

B Right   **E** Equilateral

C Obtuse

8. The set of ordered pairs, when plotted on a coordinate plane and connected in order, form what type of two-dimensional figure?

$Q(-5, -2)$, $R(6, -1)$, $S(5, 10)$, $T(-6, 9)$

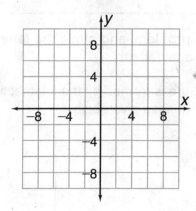

A Triangle

B Rectangle

C Circle

D Rhombus

E Trapezoid

9. The set of ordered pairs, when plotted on a coordinate plane and connected in order, form what type of two-dimensional figure?

$L(10, 14)$, $M(14, 10)$, $N(10, 6)$, $P(6, 10)$

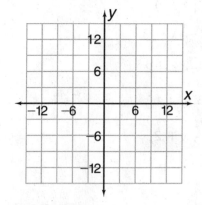

A Rhombus

B Rectangle

C Square

D Parallelogram

E All of the above

Grade 9

Name _____ Date _____ Class _____

Test Preparation Practice
Data Analysis and Probability

12.4.1.a Read or interpret data, including interpolating or extrapolating from date.

Solve each problem, and circle the letter of the best answer.

Use the graph to answer Questions 1 and 2.

A group of randomly selected students were asked to name their favorite food.

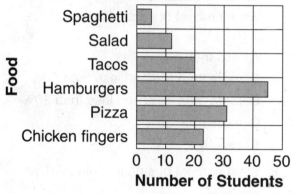

Favorite Foods

1. Estimate the number of students that participated in the survey.

   A 5

   B 45

   C 99

   D 130

   E 160

2. Approximately 5% of the students surveyed voted for which food?

   A Chicken fingers

   B Hamburgers

   C Salad

   D Spaghetti

   E Pizza

Use the graph to answer Questions 3 and 4.

The graph shows the number of cars Richard sold during a 7-month period.

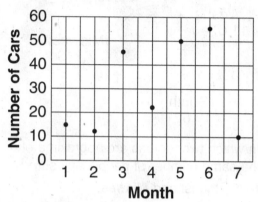

3. Which data set could be described by the line graph?

   A 10, 10, 30, 80, 90, 100, 110

   B 50, 45, 40, 30, 13, 5

   C 14, 12, 46, 20, 50, 56, 10

   D 136, 34, 34, 56, 100, 2, 8

   E 10, 15, 6, 45, 45, 10, 10

4. Between which two months did the biggest increase in sales occur?

   A Months 1 and 2

   B Months 2 and 3

   C Months 3 and 4

   D Months 5 and 6

   E Months 6 and 7

Grade 9

Use the circle graph to answer Questions 5 and 6.

The graph shows the number of people who are uninsured (without health insurance) in the United States.

Number (in thousands) of Uninsured in U.S. for the Year 2003

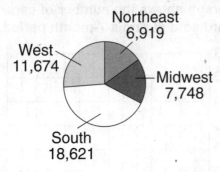

Northeast
6,919

West
11,674

Midwest
7,748

South
18,621

5. In which two regions are approximately the same number of people uninsured?

   A Northeast and Midwest

   B Midwest and South

   C South and West

   D West and Midwest

   E South and Northeast

6. How many people are uninsured in the South?

   A 6919

   B 7748

   C 18,621

   D 7,748,000

   E 18,621,000

Use the table for Questions 7–9.

The table shows the scores of a test in Professor Miller's algebra class and the percent of students with scores in each interval. A score of 70 is passing.

Score	Percent of Students
90 and above	5%
80–89	13%
70–79	20%
60–69	43%
59 and below	19%

7. What percent of students passed the test?

   A less than 5%

   B more than 5% and less than 10%

   C more than 10% and less than 20%

   D more than 35%

   E exactly 20%

8. What percent of students did NOT pass the test?

   A 5%

   B 38%

   C 40%

   D 55%

   E 62%

9. One **possible** conclusion that can be made about the exam is:

   A The students were well prepared.

   B The students were not well prepared.

   C The test was easy.

   D The teacher knew the material.

   E The students enjoyed the test.

Test Preparation Practice
Data Analysis and Probability

12.4.1.b For a given set of data, complete a graph and then solve a problem using the data in the graph (histograms, scatterplots, line graphs).

Solve each problem, and circle the letter of the best answer.

Use the scatter plot to answer Questions 1 and 2.

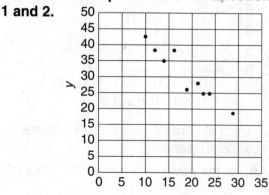

1. Which of the following graphs best describes the relationship between the variables *x* and *y*?

A
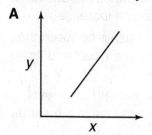

B
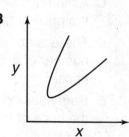

C
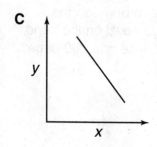

D
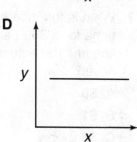

E
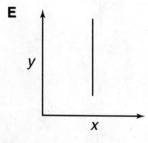

2. What is the value of *y* when *x* = 10?

   A 25

   B 35

   C 38

   D 42

   E 48

3. Which scatter plot shows a positive correlation between two variables for a set of data?

A
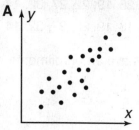

B
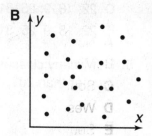

C
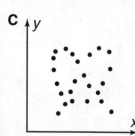

D
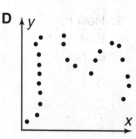

E
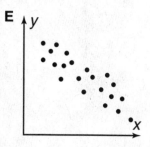

Grade 9

Use the histogram for Questions 4 and 5.

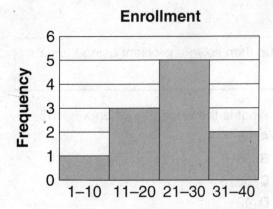

Enrollment

4. The histogram represents which data set?

  A 32, 18, 9, 25, 31, 28, 19, 22, 27, 32, 14

  B 22, 18, 9, 35, 31, 28, 19, 22, 27, 32, 14

  C 22, 8, 9, 25, 31, 28, 19, 22, 27, 32, 14

  D 22, 18, 9, 25, 31, 28, 19, 22, 27, 32, 14

  E 22, 18, 9, 25, 31, 18, 19, 22, 27, 32, 14

5. How many classes have an enrollment greater than 20?

  A 2

  B 4

  C 5

  D 3

  E 7

Use the scatter plot with its trend line to answer Questions 6 and 7.

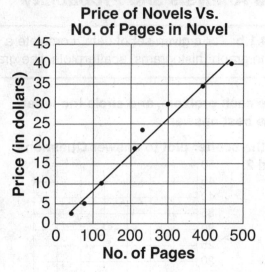

Price of Novels Vs. No. of Pages in Novel

6. Which statement is strongly supported by the scatter plot?

  A As the novel gets longer, the price of the novel decreases.

  B As the number of pages in the novel decreases, the price of the novel increases.

  C As the number of pages increases, the price of the novel increases.

  D There is no relationship between the number of pages in a novel and its price.

  E Regardless of the length of a novel, the price of the novel stays the same.

7. What is the difference between the price for a 300-page novel on the trend line and the actual price of a 300-page novel?

  A $0

  B $1

  C $3

  D $6

  E $8

Grade 9

Test Preparation Practice
Data Analysis and Probability

12.4.1.c Solve problems by estimating and computing with univariate or bivariate data (including scatter plots and two-way tables).

Solve each problem, and circle the letter of the best answer.

Use the line graph to answer Questions 1 and 2.

The graph shows the average daily temperatures for one week in Anchorage, Alaska.

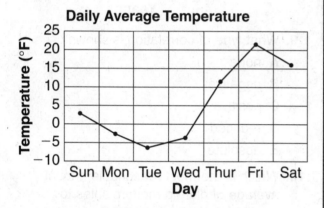

Daily Average Temperature

1. How many days had average temperatures below 0° F?

   A 0

   B 1

   C 2

   D 3

   E 4

2. Approximate the lowest daily temperature for the given week.

   A 2° F

   B 0° F

   C −3° F

   D −4° F

   E −7° F

Use the back-to-back stem-and-leaf plot for Questions 3–5.

Number of Visitors

Toniville	Stem	Ft. Humble
9 9 9	1	
1 2 3 4	2	
0 2 3 5 6	3	1 3 3 3 6 6 7 9
	4	1 1 2 2

Key: 9 | 1 means 19 and 3 | 1 means 31

3. What is the mean number of visitors to Toniville?

   A 37

   B 26

   C 36

   D 27

   E 23

4. What is the median number of visitors to Ft. Humble?

   A 36

   B 36.5

   C 37

   D 37.5

   E 33

5. What is the difference between the mode of visitors to Toniville and the mode of visitors to Ft. Humble?

   A 33

   B 19

   C 52

   D 26

   E 14

Use the line graph to answer Questions 6–8.

The line graph gives the entertainment expenses of a family for 6 weeks.

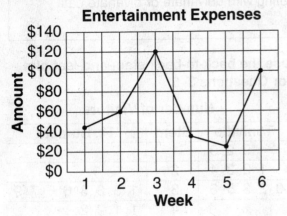

Entertainment Expenses

Use the Graph for Questions 9 and 10.

The scatter plot shows the annual number of dial-up modems sold at three different stores over a period of 5 years.

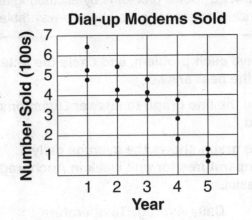

Dial-up Modems Sold

6. During which week did the family spend the most on entertainment expenses?

   A Week 1

   B Week 2

   C Week 3

   D Week 5

   E Week 6

7. Estimate the total expenses for weeks 4 and 5.

   A $20

   B $30

   C $40

   D $50

   E $60

8. What is the approximate difference between the most expensive week and the least expensive week?

   A $30

   B $50

   C $70

   D $80

   E $100

9. What type of correlation is shown?

   A none

   B negative

   C positive

   D reduced

   E absolute

10. Which is the most reasonable annual average of dial-up modem sales to predict for a store in Year 6?

   A 650

   B 500

   C 350

   D 200

   E 50

Grade 9

Name _____ Date _____ Class _____

Test Preparation Practice
Data Analysis and Probability

> **12.4.1.d** Given a graph or a set of data, determine whether information is represented effectively and appropriately (bar graphs, box plots, histograms, scatterplots, line graphs).

Solve each problem, and circle the letter of the best answer.

Use the graph to answer Questions 1 and 2.

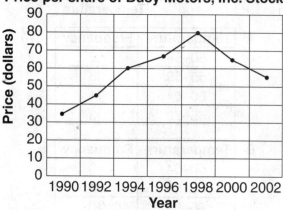

Price per share of Busy Motors, Inc. Stock

1. What is the approximate value of the stock for the year 1997?

A $67

B $70

C $72

D $75

E $80

2. During which period was the stock price increasing?

A 1990–2002

B 1996–2001

C 1990–1998

D 1994–2000

E 2000–2001

Use the pictogram to answer Questions 3 and 4.

The graph shows the average attendance at football games at Big University.

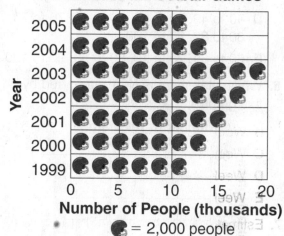

Attendance at Football Games

🏈 = 2,000 people

3. What is the difference between the greatest number of people attending and the least number of people attending?

A 8   D 8,000

B 10   E 10,000

C 6,000

4. The university received extra funding for its football program if the average attendance was over 16,000. For which year(s) did the university receive extra funding?

A 1999, 2000   D 2002, 2003

B 2001   E 2004, 2005

C 1999

Grade 9

5. The data gives the scores of an exam.

99	55	78	91	30	82	88	92	69	35
79	70	82	45	38	54	92	87		
80	34	40	30	87	92	94	48	73	

Which is a possible set of intervals for a histogram of this set of data?

A 10 to 29, 30 to 49, 50 to 79

B 20 to 29, 30 to 39, 40 to 49, 50 to 59, 60 to 69, 70 to 79, 80 to 89

C 30 to 39, 40 to 49, 50 to 59, 60 to 69, 70 to 79, 80 to 89, 90 to 100

D 40 to 49, 50 to 59, 60 to 69, 70 to 79, 80 to 89, 90 to 100

E 60 to 69, 70 to 79, 80 to 89, 90 to 100

6. Which statement is NOT supported by this graph?

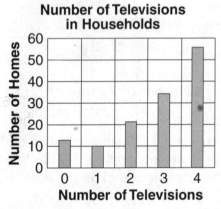

Number of Televisions in Households

A The number of households which have two televisions is more than 20.

B The number of households which have 1 television is greater than the number of households which have no televisions.

C The number of households which have 4 televisions is greater than the number of households which have 3 televisions.

D No household has 7 televisions.

E The number of households which have 3 televisions is less than 40.

7. Which set of data below is represented by the histogram?

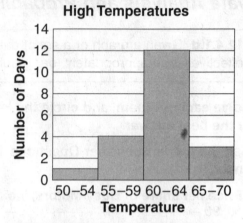

High Temperatures

A

Temperature	Frequency
50–54	2
55–59	5
60–64	11
65–70	3

B

Temperature	Frequency
50–54	0
55–59	4
60–64	10
65–70	2

C

Temperature	Frequency
50–54	1
55–59	4
60–64	14
65–70	2

D

Temperature	Frequency
50–54	1
55–59	4
60–64	12
65–70	3

E

Temperature	Frequency
50–54	3
55–59	12
60–64	4
65–70	1

Grade 9

Test Preparation Practice

Data Analysis and Probability

> **12.4.2.a** Calculate, interpret, or use mean, median, mode, range, interquartile range, or standard deviation.

Solve each problem, and circle the letter of the best answer.

Use the data to answer Questions 1–4.

The following data are the weights (in pounds) of eight competition dogs.

123, 145, 105, 98, 130, 111, 87, 105

1. Find the range of the data.
   - A 35
   - B 45
   - C 58
   - D 63
   - E 70

2. Find the median of the data.
   - A 98
   - B 105
   - C 108
   - D 110.5
   - E 114

3. Find the mode of the data.
   - A 87
   - B 98
   - C 100
   - D 105
   - E 123

4. Find the mean of the data.
   - A 98
   - B 100
   - C 110
   - D 113
   - E 130

Use the information below to answer Questions 5–7.

Sara's test scores for her anatomy class have a mean of 96 with a range of 8.

5. The sum of Sara's test scores is 576. How many tests has she taken?
   - A 3
   - B 4
   - C 5
   - D 6
   - E 7

6. Sara's highest test score is 98. What is her lowest test score?
   - A 70
   - B 75
   - C 80
   - D 85
   - E 90

7. If Sara maintains her average of 96, she will win a prize for having the highest average in the class. What is the minimum score she needs on the next test to maintain her average of 96?
   - A 92
   - B 94
   - C 96
   - D 98
   - E 100

8. Which number, if added to the data set, would affect the mean the most for the data set below?

   {11, 98, 23, 18, 29, 9, 15}
   - A 11
   - B 15
   - C 18
   - D 29
   - E 98

Name _____ Date _____ Class _____

Use the data set {15, 30, 20, 20, 35} to answer Questions 9 and 10.

9. Which statement is true about this data set?

    A The mean is 20.

    B The median is 20.

    C The mode is 30.

    D There is no mode.

    E The range is 25.

10. Which statement is NOT true about this data set?

    A The mean is 24.

    B The range is equal to the median.

    C The mode is equal to the maximum value.

    D The mode is equal to 20.

    E The mode is equal to the range.

11. Identify the measure of central tendency that is being used in this statement

    The average test score in mathematics for Chase Elementary school is 84.

    A mean   D range

    B median   E percentile

    C mode

12. Identify the measure of central tendency that is being used in this statement.

    Mitchell's salary is larger than that of half of the employees at the company where he works.

    A mean   D range

    B median   E percentile

    C mode

Use the data to answer Questions 13–16.

4 1 1 0 2 4 0 0 3 2 2 2 4

13. What is the median?

    A 0   D 3

    B 1   E 4

    C 2

14. What is the range of the data?

    A 1   D 4

    B 2   E 5

    C 3

15. Which statement is true about the data?

    A The mode is equal to the range.

    B The mode is 0.

    C The difference between the median and the mode is 0.

    D There is no mode.

    E The mean is 4.5.

16. Find the mean.

    A 0

    B 1.01

    C 1.92

    D 2.08

    E 4

Grade 9

Name _____ Date _____ Class _____

Test Preparation Practice
Data Analysis and Probability

12.4.4.b Determine the theoretical probability of simple and compound events in familiar or unfamiliar contexts.

Solve each problem, and circle the letter of the best answer.

Use the information and picture to answer Questions 1–4.

Several numbers and letters are put on cards. Bobbie hangs them on a wall, as shown below, and randomly throws a dart at one of the cards.

5	A	3	S
B	Z	8	M
N	7	2	U
E	Q	P	12

1. What is the probability of the dart landing on a vowel or on an even number?

   A $\frac{1}{5}$

   B $\frac{5}{16}$

   C $\frac{3}{8}$

   D $\frac{7}{16}$

   E $\frac{9}{14}$

2. What is the probability of the dart landing on a consonant?

   A $\frac{2}{35}$

   B $\frac{4}{25}$

   C $\frac{3}{8}$

   D $\frac{7}{16}$

   E $\frac{7}{8}$

3. What is the probability of the dart landing on an even number or a prime number?

   A $\frac{1}{16}$

   B $\frac{3}{16}$

   C $\frac{3}{8}$

   D $\frac{5}{32}$

   E $\frac{11}{18}$

4. What is the probability of the dart landing on the letter V?

   A 0

   B $\frac{3}{64}$

   C $\frac{1}{4}$

   D $\frac{9}{16}$

   E $\frac{5}{8}$

Use the information for Questions 5–7.

Each resident of a town casts one vote for chairperson of the library board. Of the residents, 35% voted for Smith, 20% voted for Masters, 25% voted for Hill, and 20% voted for Green.

5. What is the probability that a resident voted for Smith or Hill?

   A 0.15

   B 0.20

   C 0.35

   D 0.55

   E 0.60

Grade 9

6. What is the probability that a resident did NOT vote for Green?

A 0.20

B 0.45

C 0.75

D 0.80

E 0.85

7. What is the probability that a resident neither voted for Masters nor Hill?

A 0.15

B 0.20

C 0.55

D 0.70

E 0.80

Use the information for Questions 8–11.

A school has 60 teachers. Of the 40 male teachers, 2 teach history. One-twelfth of all the teachers at the school teach history.

8. What is the probability that a teacher is a man or teaches history?

A $\frac{1}{4}$

B $\frac{7}{12}$

C $\frac{41}{60}$

D $\frac{43}{60}$

E $\frac{59}{60}$

9. What is the probability that a teacher is a woman or teaches history?

A $\frac{13}{60}$

B $\frac{1}{3}$

C $\frac{11}{30}$

D $\frac{5}{6}$

E $\frac{57}{60}$

10. What is the probability that a teacher does not teach history?

A $\frac{1}{12}$

B $\frac{1}{4}$

C $\frac{1}{3}$

D $\frac{11}{12}$

E 1

11. What is the probability that a teacher does not teach history or is a man?

A $\frac{1}{20}$

B $\frac{1}{12}$

C $\frac{11}{30}$

D $\frac{7}{12}$

E $\frac{57}{60}$

12. Which pair of events is NOT mutually exclusive?

A Getting a 4 and getting an odd number when rolling one die

B Drawing a king and drawing a heart when drawing one card from a deck of playing cards

C Drawing a red card and drawing a black card when drawing one card from a deck of playing cards

D Getting a 5 and getting a 1 when rolling a die

E Selecting a green ball and selecting a black ball when selecting one ball from a crate of solid colored balls

Grade 9

Name _____ Date _____ Class _____

Test Preparation Practice
Data Analysis and Probability

12.4.4.c Given the results of an experiment or simulation, estimate the probability of simple or compound events in familiar or unfamiliar contexts.

Solve each problem, and circle the letter of the best answer.

Use the table for Questions 1–3.

Amanda tosses a fair coin and then rolls a 6-sided die several times. The table shows all possible outcomes and Amanda's frequency of getting each outcome.

Coin-Die	Frequency
H-1	3
H-2	4
H-3	3
H-4	2
H-5	1
H-6	4
T-1	3
T-2	2
T-3	3
T-4	2
T-5	4
T-6	3

1. Based on this table, what is the probability that Amanda will get heads and an even number when she tosses the coin and die?

   A $\frac{5}{34}$ **D** $\frac{5}{11}$

   B $\frac{5}{17}$ **E** $\frac{9}{17}$

   C $\frac{5}{12}$

2. What is the probability that Amanda will not get tails on her next try?

   A $\frac{1}{34}$ **D** $\frac{1}{2}$

   B $\frac{1}{5}$ **E** $\frac{21}{34}$

   C $\frac{1}{4}$

3. Which of the following is NOT true about this frequency table?

   A The probability that Amanda gets heads and the number 1 on her next try is $\frac{3}{34}$.

   B The probability that Amanda gets heads and an odd number is equal to the probability that she gets tails and an even number.

   C The probability that an outcome has heads and tails is 0.

   D The probability that Amanda gets heads and the number 6 is $\frac{1}{6}$.

   E The probability that Amanda gets tails and an odd number is greater than the probability she gets tails and the number 4.

4. A town report shows that $\frac{4}{5}$ of residents own a vehicle and $\frac{1}{2}$ of these vehicle owners own a truck. There are 3000 residents in the town. How many residents are likely to own a truck?

   A 500
   B 800
   C 1000
   D 1200
   E 2000

Grade 9

Use the table to answer Questions 5–7.

Dena collected data on the kind of sandwiches her customers order at her deli shop. The table shows the results.

Sandwich	Number of Customers
Pastrami	26
Ham	25
Veggie	15
Steak	22
Tomato/Cheese	12

5. What is the probability that the next customer will order a pastrami sandwich?

   A $\frac{1}{100}$ **D** $\frac{1}{2}$

   B $\frac{21}{100}$ **E** $\frac{67}{100}$

   C $\frac{13}{50}$

6. What is the probability that a customer will not order a steak sandwich?

   A $\frac{21}{100}$ **D** $\frac{69}{100}$

   B $\frac{22}{100}$ **E** $\frac{39}{50}$

   C $\frac{1}{4}$

7. What is the probability that a customer will order a sandwich without meat?

   A $\frac{3}{25}$ **D** $\frac{69}{100}$

   B $\frac{27}{100}$ **E** $\frac{49}{50}$

   C $\frac{37}{100}$

Use the information for Questions 8–11.
A quality inspector inspects 2000 pairs of shoes and finds 1990 to have no defects.

8. What is the probability that a pair of shoes chosen at random will have no defects?

   A 0.80   **D** 0.995

   B 0.95   **E** 1.0

   C. 0.98

9. What is the probability that a pair of shoes chosen at random will have defects?

   A 0.005

   B 0.05

   C 0.1

   D 0.3

   E 0.9

10. Predict the number of pairs of shoes that will NOT be defective when 600 shoes are produced.

    A 100

    B 488

    C 550

    D 597

    E 600

11. Four thousand pairs of shoes are shipped to a large department store. Predict the number of shoes in the shipment that are likely to have defects.

    A 5

    B 10

    C 20

    D 30

    E 40

Grade 9

Test Preparation Practice
Data Analysis and Probability

| 12.4.4.e Determine the number of ways an event can occur using tree diagrams, formulas for combinations and permutations, or other counting techniques. |

Solve each problem, and circle the letter of the best answer.

1. A menu has 3 appetizers, 4 salads, 7 entrees, and 2 desserts. How many different 4-course meals can be made from this menu?

   A 16

   B 84

   C 91

   D 168

   E 183

2. The Personal Identification Number for a bank must be a four-digit number. Each digit can be a number between zero and nine. How many possible Personal Identification Numbers are there?

   A 10   D 10,000

   B 100   E 100,000

   C 1000

3. A man tosses a six-sided die four times. How many possible outcomes are there?

   A 6   D 6^4

   B 6^2 E $6^4 - 4^6$

   C 6^3

4. A book club with 12 members needs a president, vice president, and treasurer. How many ways can the club choose the 3 officers?

   A 4

   B $_{12}P_3$

   C $_{12}C_3$

   D $_{30}C_{12}$

   E $_3P_{12}$

5. A basketball team has 15 members. How many different ways can the coach choose a captain?

   A 1

   B 3

   C 5

   D 15

   E 45

Use the spinner for Questions 6 and 7.

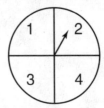

6. A boy is playing a game. He first randomly picks a red, blue, or green ball out of a bag and then spins the spinner. How many possible outcomes are there?

   A 4

   B 7

   C 12

   D 24

   E 50

7. Pat first randomly picks a red, blue, or green ball out of a bag and then spins the spinner. Which of the following is an outcome of this game?

   A Blue, 3

   B Red, Blue

   C 2, Green

   D Blue, Green

   E 1, 4

Grade 9

8. Armando wants to buy a television and DVD player. An Appliance Store has 10 different televisions and 7 different DVD players from which to choose. How many different pairs can Armando choose for a television and a DVD player?

A 7

B 10

C 35

D 70

E 100

9. A coin is tossed two times. Which tree diagram can you use to find the number of outcomes?

A

B

C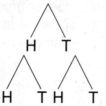

D

E

10. A stockbroker is selling 4 different annuities, 6 different mutual funds, and 8 different stocks. In how many ways can a client choose one of each type of investment for his portfolio?

A 10

B 18

C 48

D 192

E 200

11. A website requires registered users to have a password that consists of 5 letters. How many different passwords are possible if no letters may be repeated?

A 130

B 3125

C 358,800

D 7,893,600

E 165,765,600

12. A gardener needs to choose four out of eight brands of fertilizer. How many ways can he do this?

A 8

B 32

C 35

D 60

E 70

13. Chris, a radio disc jockey, has 12 songs to play during his morning show. In how many ways can he play the songs?

A 10!

B 12!

C 14!

D 16!

E 18!

Grade 9

Name _____ Date _____ Class _____

Test Preparation Practice
Data Analysis and Probability

12.4.4.f Determine the probability of the possible outcomes of an event.

Solve each problem, and circle the letter of the best answer.

Use the information for Questions 1 and 2.

Krista spins a spinner once. The spinner has six evenly spaced colors: red, yellow, pink, green, blue, and white.

1. What is the probability that when Krista spins the spinner it will NOT land on yellow?

   A $\frac{1}{6}$ **D** $\frac{2}{3}$

   B $\frac{2}{6}$ **E** $\frac{5}{6}$

   C $\frac{1}{2}$

2. Which event has a probability of $\frac{2}{3}$?

   A The spinner lands on green.

   B The spinner does not land on green or white.

   C The spinner lands on yellow or pink.

   D The spinner lands on red, yellow, and green.

   E The spinner lands on brown.

Use the information for Questions 3–6.

A fair red die and a fair green die are rolled.

3. Which of the following is the event that the sum of the dice is 7?

   A {1-6, 2-5, 3-4, 4-3, 5-2, 6-1}

   B {1-4, 1-5, 1-6, 6-1, 6-2, 6-3}

   C {1-1, 2-2, 3-3}

   D {4-4, 5-5, 6-6}

   E {1-6, 3-4, 5-2, 4-4}

4. What is the probability that a sum of 12 occurs?

   A $\frac{1}{72}$ **D** $\frac{2}{9}$

   B $\frac{1}{36}$ **E** $\frac{23}{72}$

   C $\frac{5}{36}$

5. What is the probability that sum of the numbers is a multiple of 3?

   A $\frac{1}{8}$

   B $\frac{1}{6}$

   C $\frac{1}{5}$

   D $\frac{1}{3}$

   E $\frac{1}{2}$

6. What is the probability of getting a sum less than 9?

   A 1:8 about 0.45

   B 1:9 about 0.58

   C 1:10 about 0.72

   D 1:11 about 0.75

   E 1:12 about 0.84

Grade 9

7. There is a 35% chance of rain tomorrow. What is the probability of NOT getting rain tomorrow?

A 35%

B 45%

C 55%

D 65%

E 75%

Use the information to answer Questions 8–12.

A jar has indigo, purple, green, and burgundy marbles. The probability of drawing an indigo marble is 0.40. The probability of drawing a purple marble is 0.25. The probability of drawing a burgundy marble is 0.05.

8. What is the probability of drawing a green marble?

A 0.20

B 0.30

C 0.35

D 0.40

E 0.50

9. What is the probability of NOT drawing a burgundy marble?

A 0.05

B 0.25

C 0.55

D 0.85

E 0.95

10. Which event has the probability of 0.45?

A Drawing a burgundy or indigo marble

B Drawing a green marble

C Drawing a green or indigo marble

D Drawing a yellow marble

E Drawing a burgundy, green, or indigo marble

11. Which of the following events has the greatest probability?

A Drawing a burgundy marble

B Drawing a green or indigo marble

C Not drawing an indigo marble

D Not drawing a green marble

E Not drawing a purple marble

12. Which event has about a 60% chance of NOT occurring?

A Drawing a purple marble

B Drawing a green or purple marble

C Drawing an indigo marble

D Not drawing a burgundy marble

E Not drawing a green marble

13. A fair number cube is rolled. What is the probability that a number less than 3 appears?

A $\frac{1}{6}$

B $\frac{1}{3}$

C $\frac{1}{2}$

D $\frac{2}{3}$

E $\frac{5}{6}$

Grade 9

Test Preparation Practice
Data Analysis and Probability

12.4.4.h Determine the probability of independent and dependent events.

Solve each problem, and circle the letter of the best answer.

Use the spinner to answer Questions 1–4.

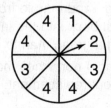

1. What is the probability of spinning a 3 once and then 3 again?

   A $\frac{1}{16}$

   B $\frac{1}{9}$

   C $\frac{1}{3}$

   D $\frac{1}{2}$

   E $\frac{3}{4}$

2. What is the probability of spinning a 1, then a 3, and then a 2?

   A $\frac{1}{300}$ D $\frac{1}{8}$

   B $\frac{1}{256}$ E $\frac{7}{8}$

   C $\frac{1}{25}$

3. What is the probability of NOT getting a 4 and then getting a 4?

   A $\frac{1}{8}$ D $\frac{1}{2}$

   B $\frac{1}{4}$ E $\frac{9}{8}$

   C $\frac{3}{8}$

4. What is the probability of getting a 1 and then NOT getting an even number?

   A $\frac{1}{64}$

   B $\frac{3}{64}$

   C $\frac{5}{64}$

   D $\frac{1}{4}$

   E $\frac{5}{8}$

Use the information to answer Questions 5 and 6.

A bag contains 20 marbles: 10 are brown and 10 are white. A marble is chosen at random and is NOT put back into the bag. A second marble is then chosen at random.

5. What is the probability of selecting a brown marble and then a white marble?

   A $\frac{1}{4}$

   B $\frac{5}{19}$

   C $\frac{2}{5}$

   D $\frac{11}{20}$

   E $\frac{3}{4}$

6. What is the probability of selecting two brown marbles in a row?

   A $\frac{3}{38}$ D $\frac{11}{38}$

   B $\frac{1}{4}$ E $\frac{4}{5}$

   C $\frac{9}{38}$

Grade 9

Use the information to answer Questions 7–9.

Two fair number cubes are rolled. One cube is green and the other cube is blue.

7. What is the probability that the product of the two numbers is less than 13 and the blue cube is also a 4?

   A $\frac{1}{16}$ D $\frac{1}{2}$

   B $\frac{1}{12}$ E $\frac{3}{4}$

   C $\frac{5}{36}$

8. What is the probability of getting a sum of 5 and the green cube is less than 5?

   A $\frac{1}{72}$ D $\frac{1}{6}$

   B $\frac{1}{36}$ E $\frac{41}{72}$

   C $\frac{1}{9}$

9. Which of the following statements is NOT true?

   A The probability that the sum of the two cubes is less than 10 and the blue cube shows a 4 is $\frac{5}{36}$.

   B. The probability that the product of the two cubes is less than 20 and the blue cube shows a 6 is $\frac{1}{12}$.

   C The probability that the sum of the two cubes is less than 5 and the blue cube shows an 1 is $\frac{5}{36}$.

   D The probability that the blue cube shows a 6 and the green cube shows a 6 is $\frac{1}{6}$.

   E. The probability that the blue cube shows a multiple of 3 and the green cube shows a 4 is $\frac{1}{18}$.

10. Which pair of events is independent?

    A Rolling a die and then drawing a card from a deck of playing cards

    B Drawing two aces from a deck of 52 playing cards when the first card is not replaced

    C Drawing two odd-numbered tickets from a jar when the tickets are drawn without replacement

    D Drawing a red marble from a jar, not putting it back into the jar, and then drawing a green marble

    E Selecting 4 non white marbles from a jar without replacement

11. The spinner shown is spun once, and a coin is flipped once.

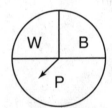

What is the probability of the coin landing with heads up and the spinner landing on *P*?

   A $\frac{1}{5}$

   B $\frac{1}{4}$

   C $\frac{1}{2}$

   D $\frac{3}{4}$

   E $\frac{5}{6}$

Grade 9

Test Preparation Practice
Algebra

12.5.1.a Recognize, describe, or extend arithmetic, geometric progressions or, patterns using words or symbols.

Solve each problem, and circle the letter of the best answer.

1. Which of the following would be considered a pattern?

   A ☼☼✈☼🖐☼☼☼✈✈☼☼☼☼☼✈ ☼🖐☼☼✈✈☼☼

   B ☼✈✈🖐☼☼✈✈🖐☼☼✈✈🖐☼ ✈✈🖐

   C 🖐🖐☼✈🖐🖐✈☼☼🖐☼☼☼🖐 🖐☼✈🖐🖐✈

   D ☼✈☼✈🖐🖐☼✈☼✈☼✈☼ ✈☼✈☼✈

   E ✈✈✈☼🖐🖐🖐☼✈✈🖐🖐☼ ✈☼✈🖐🖐

2. What number appears in the 4th position of the given pattern?

   2 4 6 8 9 2 4 6 8 9 2 4 6 8 9 2 4 6 8 9

   A 9

   B 4

   C 2

   D 6

   E 8

3. A number that repeats after the decimal point is a rational number. A number that has a pattern but does not repeat after the decimal point is an irrational number. Which of the following is an irrational number?

   A 3.33333 . . .

   B 3.03030303 . . .

   C 3.13113111311113

   D 3.0$\overline{27}$

   E 3.1616161616

4. Which of the following number sequences has a pattern?

   A 34 41 48 55 62

   B 24 26 23 25 24

   C 20 19 22 25 28

   D 15 19 23 21 20

   E 36 38 40 39 37

5. Describe how the terms change in the following sequence.

   5, 7, 11, 17, 25, . . .

   A The terms increase by 2.

   B The terms increase by a power of 2.

   C The terms increase by an odd number.

   D The terms increase by 2, then increase by 4, increase by 2, then 4.

   E The terms increase by consecutive even numbers.

6. Describe the operation performed on one term to equal the next term?

   600, 300, 150, 75, ...

   A Divide by 300.

   B Divide by 2.

   C Subtract 300.

   D Subtract a power of 3.

   E Divide by a power of 3.

7. Which of the following represents an arithmetic sequence?

   A 1250, 2500, 3500, 4750, . . .

   B 8, 13, 18, 23, 28, . . .

   C 1, 2, 3, 1, 2, 3, . . .

   D 5, 9, 7, 11, 15, 13, . . .

   E 5, 1, -3, -5, -7, . . .

Grade 9

8. Which of the following is considered a pattern?

A △△▲■□▽▼△▲■□
△▲■□△▲▲■□

B △▲■□○●△▲■□○
△▲■□○△▲▲■□○

C △▲■□▽▼△▲▲■□▽▼
△▲■□▽▼△▲■□▽▼

D □○●▽▼□○●△▲
□●▽□○●▽▼

E △▲■□△▲▲△▲■
△▲■□△△▲■□△▲

9. What is the pattern in the sequence?

96, 48, 24, 12, 6, . . .

A Subtract 48.

B Divide by 2.

C Divide by a power of 2.

D Multiply by 0.75.

E Divide by 0.5.

10. What is the pattern in the sequence?

6, 18, 54, 162, . . .

A Divide by 3.

B Divide by 0.3.

C Multiply by 0.3.

D Multiply by a power of 3.

E Multiply by 3.

11. What is the pattern in the sequence?

6.5, 42.25, 274.625, 1785.0625, . . .

A Increase by a power of 6.5.

B Multiply by 6.2.

C Add 37.75.

D Multiply by 3.375.

E Multiply by a power of 3.

12. The ordered pairs shown form a quadratic pattern.

x	y
0	1
1	2
2	5
3	10
4	17
5	??

What is the missing *y*-value?

A 10

B 22

C 24

D 26

E 28

13. What shape appears in the 3rd position of the pattern?

A ✚

B ⬠

C ☺

D ⌒

E ♡

14. As the *x*-values increase by 4, the *y*-values:

x	−6	−2	2	6	10
y	12	9	6	3	0

A increase by $\frac{1}{2}$.

B increase by 2.

C decrease by $\frac{1}{2}$.

D decrease by 3.

E decrease by −3.

Grade 9

Test Preparation Practice
Algebra

12.5.1.b Express the function in general terms (either recursively or explicitly), given a table, verbal description, or some terms of a sequence.

Solve each problem, and circle the letter of the best answer.

Use the table for Questions 1 and 2.

C°	0	10	20	30	40
F°	32	50	68	86	104

1. Which describes the pattern in the table?

   A The Celsius temperatures increase by 18° for every increase of 10° in the Fahrenheit temperatures.

   B The Fahrenheit temperatures are 32° more than the Celsius temperatures.

   C The Celsius temperatures increase by 10° and the Fahrenheit temperatures have no pattern.

   D The Fahrenheit temperatures increase by 18° for every increase of 10° in the Celsius temperatures.

   E The Celsius temperatures are about 40° less than the Fahrenheit temperatures.

2. What Fahrenheit temperature corresponds to a Celsius temperature of 50°?

   A 144°   **D** 82°

   B 122°   **E** 158°

   C 140°

3. Which function best describes the pattern shown in the table?

x	1	2	3	4	5
y	1	8	27	64	125

   A $y = 4x$ **D** $y = 9x$

   B $y = x^2$ **E** $y = x + 3x$

   C $y = x^3$

Use the dot pattern for Questions 4–6.

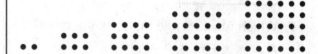

4. Which best describes the dot pattern?

   A The number of dots in each rectangle are multiples of 6.

   B The number of dots in each rectangle doubles from left to right.

   C Each rectangle from left to right has one more row of dots and one more column of dots than the previous rectangle.

   D The diagonal of each rectangle from left to right contains one more dot than the previous rectangle.

   E The rectangles are all similar to a 1 unit by 2 unit rectangle.

5. What number sequence best represents the dot pattern?

   A 1, 2 3, 4, 5

   B 1, 4, 9, 16, 25

   C 2, 6, 18, 54, 162

   D 2, 6, 12, 20, 30

   E 2, 3, 4, 5, 6

6. What expression for a_n can be used to complete the table representing the dot pattern?

n	1	2	3	4	...	n
a_n	2	6	12	20	...	?

   A n^2 **D** $n(n + 1)$

   B $n + n + 1$ **E** $2n$

   C $n \cdot n + 1$

Grade 9

Use the table for Questions 7 and 8.

x	y
3	9
6	18
9	27
12	36

7. Which function rule matches the table?

  A $y = x + 6$

  B $y = 2x + 3$

  C $y = 3x - 1$

  D $y = 3x$

  E $y = \frac{1}{3}x$

8. What y-value corresponds to an x-value of 7?

  A $\frac{7}{3}$ D 21

  B 13   E 28

  C 16

9. The area of a shaded circle is about 28 cm^3. When the circle is inscribed in a square, the area of the white region is 8 cm^2. When two squares are placed in a rectangle as shown, the area of the white region is 16 cm^2. Using the pattern identified in the diagram, what will be the area of the white region for 6 circles?

  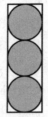

$A_w = 8$ cm^2 $A_w = 16$ cm^2 $A_w = 24$ cm^2

  A 26 cm^2

  B 36 cm^2

  C 48 cm^2

  D 144 cm^2

  E 168 cm^2

Use the table for Questions 10 and 11.

People	1	2	3	4	5
Conversations	0	1	3	6	10

10. Which describes the pattern in the table?

  A The number of conversations are consecutive prime numbers.

  B The number of conversations is half the number of people.

  C As the number of people increases by 1, the number of conversations increases by 1 more than the previous difference.

  D As the number of conversations increases, the number of people decreases.

  E As the number of people increases by 1, the number of conversations doubles.

11. What number of conversations corresponds to 10 people?

  A 36

  B 55

  C 50

  D 19

  E 45

Grade 9

Test Preparation Practice
Algebra

12.5.1.g Determine the domain and range of functions given various contexts.

Solve each problem, and circle the letter of the best answer.

1. In the graph below, what range elements correspond to domain elements {1, 2, 4}?

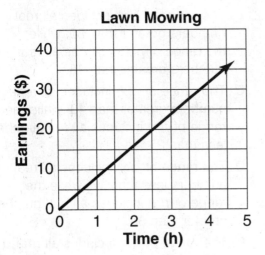

Lawn Mowing

A {1, 2, 4}

B {8, 16, 32}

C Time

D {4, 8, 12}

E Earnings

2. The side length of a right triangle is represented by the function

$f(x) = \sqrt{(x^2 - 36)}$, where x is the length of the hypotenuse. Which value is NOT in the reasonable domain of this function?

A −6.2   **D** 7

B 12.5   **E** $\frac{47}{5}$

C $8\frac{1}{4}$

3. Which best represents the domain and range of the following situation?

A person in a helicopter flies at a constant height for awhile, then dives into the sea and swims at a constant depth for awhile.

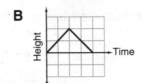
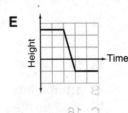
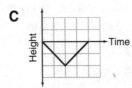

4. The path of a projectile is modeled by the equation $h = -16t^2 + 48t$, where t is time in seconds and h is height in feet. What value is NOT in the range of this function?

A 0 feet

B 10 feet

C 20 feet

D 30 feet

E 40 feet

Grade 9

5. What is the range of the function graphed?

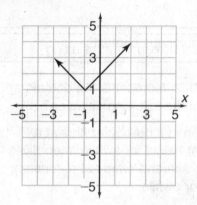

A All real numbers between 1 and 4

B All real number less than 4

C All real numbers greater than or equal to −1

D All real numbers greater than or equal to 1

E All real numbers between −3 and 2

6. The graph below shows the horsepower needed as the revolutions per minute of an agitator increases. What is the most valid statement about this function?

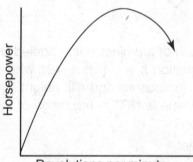

A The domain includes both positive and negative values.

B The range includes both positive and negative values.

C For at least one x-value there is more than one y-value.

D For at least one y-value there is more than one x-value.

E The domain includes the set of all real numbers.

7. The linear function $f(x) = 25x + 30$ represents a consultant's fees and the linear function $g(x) = 40x + 18$ represents another consultant's fees. If x represents the time a consultant works, how do the two reasonable ranges compare?

A The range of $g(x)$ includes all real numbers greater than 40, while the range of $f(x)$ includes all real numbers greater than 25.

B The range of $f(x)$ includes all real numbers greater than 30, while the range of $g(x)$ includes all real numbers greater than 18.

C The range of $g(x)$ includes all real numbers greater than 40, while the range of $f(x)$ includes all real numbers less than 25.

D The range of $g(x)$ includes all real numbers less than 18, while the range of $f(x)$ includes all real numbers greater than 30.

E The range of $f(x)$ includes all real numbers less than 40, while the range of $g(x)$ includes all real numbers less than 25.

8. A function is modeled by the equation $y = \frac{1}{2}x^2 + 5$. Which value is NOT in the range of this function?

A 5.3

B 5

C 3

D $8\frac{2}{3}$

E $|-9|$

Grade 9

Name _____ Date _____ Class _____

Test Preparation Practice
Algebra

12.5.2.a Translate between different representations of algebraic expressions using symbols, graphs, tables, diagrams, or written descriptions.

Solve each problem, and circle the letter of the best answer.

1. Which of the following is the correct translation of "9 less than p"?

   A $p + 9$

   B $9 + p$

   C $p - 9$

   D $9 - p$

   E $\dfrac{9}{p}$

2. Which expression represents the verbal phrase "the sum of three times a number and five"?

   A $3(n + 5)$

   B $3 + n \times 5$

   C $3(5) + n$

   D $3 + (n + 5)$

   E $3n + 5$

3. Solve the equation for x.

   $4x = -32x - 3$

   A $x = -12$

   B $x = -\dfrac{1}{12}$

   C $x = \dfrac{1}{12}$

   D $x = 2$

   E $x = 12$

4. Looking at the table, which equation best describes the relationship between the number of students and the number of tables in the cafeteria?

Number of students, (n)	Number of tables in cafeteria, (t)
720	18
600	15
960	24

   A $t = 40n$

   B $n = 40t$

   C $n = 35t + 90$

   D $t = 35n + 90$

   E $n = 45t - 90$

5. Which expression is equivalent to $7x - 42$?

   A $7(x - 42)$

   B $7(x + 42)$

   C $6(x - 7)$

   D $\dfrac{7}{x} - 42$

   E $7(x - 6)$

6. The maximum grade on a test is 85. The minimum grade is 60. If n represents a grade, which sentence best expresses this situation?

   A $85 = n = 60$

   B $85 \leq n \leq 60$

   C $60 \leq n \leq 85$

   D $85 \leq n \geq 60$

   E $60 \geq n \geq 85$

Grade 9

7. Hector has 6 times as many coins as Wilma. If Hector has 126 coins, which equation would you use to find out how many coins Wilma has?

A $c + 6 = 126$

B $6c = 126$

C $6(126) = c$

D $c = \dfrac{126 + 6}{6}$

E $126c = 6$

8. Which value of h makes the equation true?

$$\dfrac{3}{4}(24 - 8h) = 2(5h + 1)$$

A $h = 1$

B $h = 2\dfrac{3}{4}$

C $h = 3$

D $h = \dfrac{17}{4}$

E $h = 4$

9. A swimming pool charges a $75 membership fee per year, and $1.50 each time you bring a guest. Which equation shows the yearly cost y in terms of the number of guests g?

A $y = 75g + 1.5$

B $y = -1.5g + 75$

C $y = 1.5g + 75$

D $y = 1.5g + 75g$

E $y = 75 - 1.5g$

10. Patsy writes an expression that will always represent an odd number. Which expression did she write?

A $m + 1$

B $2m + 1$

C m^2

D $m^2 + 1$

E $m^2 + 2$

11. The cost of a long distance call is $2.50 for the first three minutes plus $0.10 for each additional minute. Theo called long distance and talked for 8 minutes. Which equation would best be used to find the cost, c, for the call?

A $\$2.50 \times 0.10 = c$

B $\$2.50 + c = 8$

C $\$2.50 + 0.10 = c$

D $\$2.50 + (5 \times 0.10) = c$

E $(5)(0.10) = \$2.50c$

12. A computer store receives a delivery of 512 printers. Now there are 2115 printers in stock. Which equation would you use to determine how many printers the store had before the delivery?

A $2115 + 512 = p$

B $p = 215 - 2115$

C $p + 512 = 2115$

D $512n = 2115$

E $\dfrac{512}{2115} = n$

13. Which equation describes the graph below?

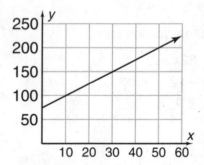

A $y = \dfrac{5}{2}x + 75$

B $y = 2x + 50$

C $y = 2x + 75$

D $y = \dfrac{5}{2}x + 50$

E $y = \dfrac{1}{2}x + 75$

Grade 9

Test Preparation Practice
Algebra

> **12.5.2.c** Graph or interpret points that are represented by ordered pairs of numbers on a rectangular coordinate system.

Solve each problem, and circle the letter of the best answer.

Use the graph below for Questions 1–4.

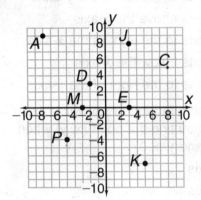

1. What are the coordinates of point *P*?

A (5, −4) **D** (−5, −4)

B (5, 4) **E** (0, −5)

C (−5, 4)

2. In which quadrant or on which axis is point *D*?

A I **D** IV

B II **E** *y*-axis

C III

3. The ordered pair (−3, 0) is represented by which point?

A *C* **D** *J*

B *D* **E** *M*

C *E*

4. For which point is *x* > 4 and *y* ≤ −2?

A *C* **D** *K*

B *E* **E** *P*

C *M*

5. In which quadrant or on which axis would you find the point (6, −16)?

A I

B II

C III

D IV

E *y*-axis

6. Which of the following list of ordered pairs do NOT lie on the same line?

A (5.5, 3), (1.5, 3), (−4, 3)

B (1, 2), (0, 0), (−3, −4)

C (−1, 0), (2, 1), (5, 2)

D (5, −2), (2, 1), (1, 2)

E (1, 1), (3, 7), (5, 13)

7. In which quadrant or on which axis would you find the point (0, −2)?

A *y*-axis

B *x*-axis

C I

D II

E IV

Grade 9

Use the graph for Questions 8–10.

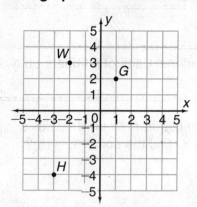

8. Which is the ordered pair for point *J*?

   A (5, 0)

   B (5, 5)

   C (−5, 1)

   D (−5, 0)

   E (0, −5)

9. What is the difference between the *y*-coordinate of *H* and the *y*-coordinate of *R*?

   A 2

   B 3

   C 8

   D −3

   E −5

10. The sum of the coordinates of which point is −5?

   A *G*

   B *H*

   C *J*

   D *R*

   E *W*

11. Which ordered pair lies in Quadrant II on a coordinate plane?

   A (−2, 2)

   B (0, 4)

   C (−2, −4)

   D (4, 2)

   E (4, −2)

12. Which of the following describes the ordered pair (−3, *y*)?

   A 3 units to the left of the *y*-axis.

   B 3 units above the *x*-axis

   C 3 units to the right of the *x*-axis

   D 3 units below the *x*-axis

   E 3 units to the right of the *y*-axis.

13. In which quadrant or on which axis would you find the point (*x*, *y*) when $x < 0$ and $y > 0$?

   A *x*-axis

   B *y*-axis

   C IV

   D II

   E III

Test Preparation Practice
Algebra

12.5.2.d Perform or interpret transformations on the graphs of linear and quadratic functions.

Solve each problem, and circle the letter of the best answer.

1. Compared with its parent function, the graph of the function $y = x^2 - 3$ is:

   A translated upward 3 units.

   B translated downward 3 units.

   C translated right 3 units.

   D translated left 3 units.

   E reflected 3 units.

2. Which of the following can be used to identify the amount and direction of the horizontal translation given by $y = a(x - h)^2 + k$?

   A a

   B x

   C h

   D y

   E k

3. Which function results when the parent function $y = x^2$ is compressed vertically by a factor of $\frac{1}{2}$?

   A $y = 2x^2$

   B $y = \frac{1}{2}x^2$

   C $y = 4x^2$

   D $y = \frac{1}{4}x^2$

   E $y = -\frac{1}{2}x^2$

4. Which is the parent function for $y = -2(x + 3)^2 - 1$?

   A $y = x + 3$

   B $y = 2x$

   C $y = x$

   D $y = 2^x$

   E $y = x^2$

5. Which function results when the parent function $y = x^2$ is compressed horizontally by a factor of $\frac{1}{2}$ and reflected across the *x*-axis?

   A $y = -\frac{1}{2}x^2$

   B $y = -\left(\frac{1}{2}x\right)^2$

   C $y = \left(-\frac{1}{2}x\right)^2$

   D $y = -2x^2$

   E $y = (-2x)^2$

6. What type of transformation is applied to the function $y = x$ to obtain the function $y = 3 - x$?

   A reflection and translation

   B reflection and compression

   C rotation and translation

   D reflection and rotation

   E stretch and rotation

Grade 9

7. What is the transformation on the parent function $y = |x|$ to obtain the graph below?

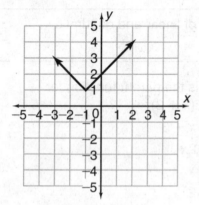

A translation 1 unit left and 1 unit down

B translation 1 unit up and 1 unit left

C translation 1 unit up and compression

D reflection across the y-axis and translation 1 unit left

E stretch and translation 1 unit up

8. The point (3, 9) is on the graph of the parent function $y = x^2$. What are the coordinates of the corresponding point on the graph of $y = -x^2 + 3$?

A (3, 12)

B (3, −12)

C (3, 78)

D (3, −84)

E (9, −84)

9. In the function $y = -4(x + 3)^2 - 6$, which transformation is NOT included on the parent function $y = x^2$?

A horizontal compression

B vertical reflection

C vertical stretch

D horizontal translation

E vertical translation

10. What transformation of the parent function is the graph of $y = -x^2$?

A Translation

B Rotation

C Reflection

D Stretch

E Compression

11. Which transformation will change the point (4, 8) on the graph of $y = 4x$ to the point (8, −8) on the graph of the transformed function?

A Translation of 4 units to the right and a reflection across the x-axis

B Reflection across the y-axis and translation of 4 units to the left

C Reflection across the x-axis and translation of 4 units to the left

D Reflection across the x-axis and a vertical stretch by a factor of 2

E Reflection across the y-axis and a vertical stretch by a factor of 2

Grade 9

Test Preparation Practice
Algebra

12.5.2.d Perform or interpret transformations on the graphs of linear and quadratic functions.

Solve each problem, and circle the letter of the best answer.

1. Which of the following is the equation of $y = -2x + 3$ shifted 2 units to the left and 4 units down?

   A $y = -2x + 4$

   B $y = -2x - 5$

   C $y = 2x - 5$

   D $y = 2x + 6$

   E $y = -2x - 8$

2. What is the change in the graph of $f(x) = (x - 1)^3 - 2$ if the function changed to $f(x) = (x + 3)^3 + 3$?

   A The graph shifts 3 units left and 3 units up.

   B The graph shifts 3 units right and 3 units down.

   C The graph shifts 3 units right and 5 units up.

   D The graph shifts 4 units left and 5 units up.

   E The graph shifts 4 units right and 5 units down.

Use the graph of $y = |x|$ to answer Questions 3 and 4.

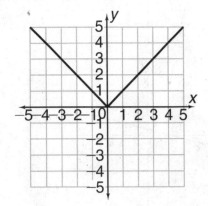

3. If the graph is shifted 2 units left, what is the y-intercept of the shifted graph?

   A $(0, 1)$

   B $(0, 2)$

   C $(0, 3)$

   D $(0, 4)$

   E $(0, 5)$

4. If the graph is shifted 2 units right and 4 units up, what is the equation of the shifted graph?

   A $y = |x + 2| - 4$

   B $y = |x - 2| + 4$

   C $y = |x - 4| + 2$

   D $y = |x - 4|$

   E $y = |x + 2| + 4$

5. The transformation from the graph of $f(x) = 2x^2 + 4$ to the graph of $g(x) = 2x^2$ is:

   A a shift 4 units up

   B a shift 4 units down

   C a shift 4 units right and 3 units down

   D a shift 4 units left

   E a shift 4 units right and 2 units up

Grade 9

6. Which of the following graphs is the graph of $f(x) = 0.25x - 1$ shifted 4 units to the right and 3 units down?

A

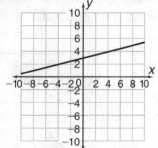

B

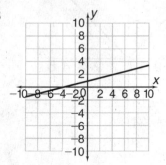

C

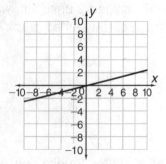

D

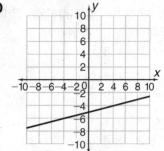

E

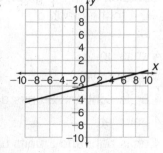

7. The functions f and g are graphed.

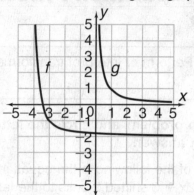

What is the relationship between $f(x)$ and $g(x)$?

A $f(x) = g(x) - 4$

B $f(x) = g(x + 4) - 2$

C $f(x) = g(x - 4) + 2$

D $f(x) = g(x - 2) - 4$

E $f(x) = g(x + 4)$

8. The graph of $g(x) = 2x + 3$ is a translation 4 units up and 1 unit right from the graph of which function?

A $g(x) = 2x - 4$

B $g(x) = 2x + 5$

C $g(x) = 2x + 9$

D $g(x) = 2x - 3$

E $g(x) = 2x + 1$

Grade 9

Name _____ Date _____ Class _____

Test Preparation Practice
Algebra

12.5.3.b Write algebraic expressions, equations, or inequalities to represent a situation.

Solve each problem, and circle the letter of the best answer.

1. Janie has 3 less than 4 times the number of songs on her MP3 player than Paulina. If p represents the number of songs Paulina has, which expression represents the number of songs Janie has?

   A $3 - 4p$

   B $3p - 4$

   C $4p + 3$

   D $4 - 3p$

   E $4p - 3$

2. The side lengths of a rectangle are $2.5x$ cm and $5.25x - 6$ cm. The area of the rectangle shown is 38 cm². Which equation can be used to determine the length of the rectangle?

   A $2(2.5x)(5.25x - 6) = 38$

   B $2(2.5x) + 2(5.25x - 6) = 38$

   C $\dfrac{(2.5x) + (5.25x - 6)}{2} = 38$

   D $(2.5x)(5.25x - 6) = 38$

   E $(2.5x)(5.25x - 6) = 38^2$

3. Gil's golf coach wants him to have an average score of -5 (or 5 under par). So far Gil has earned the following scores in his first 7 golf matches: -4, -5, -12, -3, -8, -9, and -2. Which equation can be used to find the score that Gil needs next in order for his average to be -5?

   A $-43 + x = 40$

   B $43 + x = -40$

   C $\dfrac{-43 + x}{8} = -5$

   D $\dfrac{43 + x}{8} = -5$

   E $\dfrac{-43 + x}{7} = -5$

4. An electric car sells for about $25,000. If Evan purchases this car, his monthly electric bill will increase about $25. Evan currently spends about $300 per month for gasoline but has no monthly payments for his current car. Which inequality can be used to find the number of months Evan will have to own the electric car in order for it to be less expensive than his current car?

   A $25{,}000 - 25m < 300m$

   B $300m - 25m < 25{,}000$

   C $300m + 25{,}000 > 25m$

   D $25{,}000 < 25m - 300m$

   E $25{,}000 + 25m < 300m$

Grade 9

5. Which inequality can be used to determine the three greatest consecutive even integers whose sum is less than 102?

   A $3x < 102$

   B $3x + 8 < 102$

   C $x + (x + 1) + (x + 2) < 102$

   D $x + (x + 2) + (x + 4) \leq 102$

   E $x + (x + 2) + (x + 4) < 102$

6. Which equation can be used to find the unknown angle measure?

   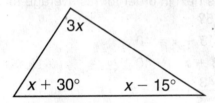

   A $5x + 15 = 90$

   B $5x + 45 = 180$

   C $3x^3 + 15 = 180$

   D $5x + 15 = 180$

   E $3x + 30 = 180$

7. The sum of two numbers is no more than 25. Twelve less than 4 times one of the numbers is 16 more than twice the other number. Which inequality can be used to determine one of the numbers?

   A $\frac{3}{2}y + 7 \geq 25$

   B $3x - 14 \leq 25$

   C $2x - 14 \leq 25$

   D $\frac{3}{2}y + 7 \geq 25$

   E $\frac{1}{2}y + 7 \leq 25$

8. Which situation correctly represents the equation $r_B = \frac{1}{2}r_A + 6$?

   A The radius of circle A is 6 less than $\frac{1}{2}$ of circle B.

   B The radius of circle A is 6 more than twice the radius of circle B.

   C The radius of circle A is 6 more than $\frac{1}{2}$ of circle B.

   D The radius of circle A is $\frac{1}{2}$ the radius of circle B.

   E The radius of circle A is 6 more than the radius of another circle.

9. Let ℓ represent the length and w represent the width of a rectangle. Which choice best describes the meaning of the equation $\ell = 10 + 4w$?

   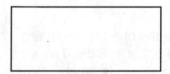

   A The length of the rectangle is 10 less than 4 times the width.

   B The length of the rectangle is 10 more than 4 times the width.

   C The length of the rectangle is 4 more than 10 times the width.

   D The width of the rectangle is 10 more than 4 times the length.

   E The length of the rectangle is 4 less than 10 times the width.

Grade 9

Test Preparation Practice
Algebra

12.5.3.c Perform basic operations, using appropriate tools, on algebraic expressions (including grouping and order of multiple operations involving basic operations, exponents, roots, simplifying, and expanding).

Solve each problem, and circle the letter of the best answer.

1. The cost to mail a first class letter is $0.37 for the first ounce plus $0.23 for each additional ounce or part of an ounce. How much does it cost to send a letter that weighs 11.4 ounces?

   A $1.00

   B $2.37

   C $2.67

   D $2.90

   E $3.10

2. Mr. Sanchez needs to fence a rectangular plot of land for his garden. How much fencing does he need?

   25 yards

   10 yards

   A 35 yards

   B 45 yards

   C 50 yards

   D 70 yards

   E 75 yards

3. Lucinda needs to buy 12 pairs of children's boots. One pair of boots costs $29.50. How much will she spend in total?

   A $295

   B $314.50

   C $325.50

   D $354

   E $360.50

4. Christa spent $42.79 total on a table cloth and sales tax. The state sales tax is 7%. What was the price of the table cloth?

   A $29.99

   B $34.99

   C $35.99

   D $39.99

   E $40.99

5. A barge traveled 245 kilometers at a rate of 7 km/hour. How long did it take the barge to travel that distance?

   A 20 hours

   B 30 hours

   C 35 hours

   D 40 hours

   E 42 hours

6. The page numbers of the two facing pages in the picture add up to 119. What page number is x?

   A 56

   D 59

   B 57

   E 60

   C 58

Grade 9

Use the information to answer Questions 7 and 8.

A-Great Copy Center sent Chelie the following invoice for her latest order.

No. of Copies	Price per Copy	Subtotal
c	$0.05	x
Processing fee	$ 3.00
Total	$15.80

7. Which expression is equal to x?

    A $15.80 ÷ $3.00

    B $15.80 − $3.00

    C $15.80

    D $15.80 + $3.00

    E $15.80 × $3.00

8. Find the value of c.

    A 132   **D** 256

    B 157   **E** 316

    C 232

9. Evaluate $x^3 − 2x^2 + 5x − 10$ for $x = −3$.

    A −80   **D** −69

    B −75   **E** −65

    C −70

10. Johann needs to cut a piece of wood that is 25 feet long into two pieces. One piece has to be 3 feet longer than the other. What is the length of the shorter piece?

    A 10 feet

    B 11 feet

    C 12 feet

    D 13 feet

    E 14 feet

11. Which value of r makes the following equation true?

    $r ÷ 5 = 33$

    A 100   **D** 180

    B 130   **E** 200

    C 165

12. Amber is asked to simplify the following expression.

    $4t^3 − 5t + 12t^2 − 12 − (3t^4 + t^2 − 12t + 15)$

    After she finishes simplifying, what is the coefficient of the t term?

    A 17

    B −7

    C −5

    D 7

    E 12

13. The local elementary school can spend at most $550 for its sports awards banquet at a neighboring restaurant. The restaurant charges a $30 setup fee plus $8 per person. What is the maximum number of people that can attend the banquet?

    A 35   **D** 65

    B 45   **E** 75

    C 55

14. Simplify: $\dfrac{3x\sqrt{6^2 + 8^2}}{5x}$.

    A 6   **D** $30x$

    B $10x$ **E** $48x$

    C 14

15. Marian's test scores on the first four tests of her biology course are 75, 88, 47, and 76. She needs an average of 70 in order to pass the course. What is the least she can score on the next test in order to pass the course?

    A 59   **D** 81

    B 64   **E** 90

    C 72

Grade 9

Test Preparation Practice
Algebra

12.5.3.d Write equivalent forms of algebraic expressions, equations, or inequalities to represent and explain mathematical relationships.

Solve each problem, and circle the letter of the best answer.

1. Which inequality is equivalent to $x \leq -4$?

  A $x \geq \dfrac{1}{4}$

  B $-x \leq -4$

  C $-x \geq 4$

  D $-x \leq \dfrac{1}{4}$

  E $-x \geq \dfrac{-1}{4}$

2. The expression

  $x \cdot x^2 \cdot x^3 \cdot x^4$

  is equal to:

  A x^{10}

  B $(x + x + x + x)^{10}$

  C x^{24}

  D x^9

  E $4x^9$

3. What are the values of m and n in the matrices?

$$\begin{bmatrix} 6 & -2 \\ m & 0 \end{bmatrix} = \begin{bmatrix} n & -2 \\ -8 & 0 \end{bmatrix}$$

  A $m = 6, n = -2$

  B $m = -8, n = 6$

  C $m = 0, n = -2$

  D $m = -8, n = -2$

  E $m = 6, n = -8$

4. What is the value of y in the matrix below?

$$\begin{bmatrix} -12 & 10 & -11 \\ 10 & 11 & 12 \end{bmatrix} + \begin{bmatrix} 8 & 6 & -5 \\ -8 & 6 & 5 \end{bmatrix} = \begin{bmatrix} 4 & y & -16 \\ 2 & 17 & 17 \end{bmatrix}$$

  A $y = -16$ **D** $y = 16$

  B $y = -6$ **E** $y = 17$

  C $y = 4$

5. Which of the following is equivalent to $(8x - 4z) - (5z + x)$?

  A $8x + 4z + 5z - x$

  B $8x + x - 4z - 5z$

  C $8x - 4z + 5z - x$

  D $9(x - z)$

  E $7x - 9z$

6. Which of the following is equivalent to $2x(3x - 1)$?

  A $2(3x^2 - x)$

  B $4x$

  C $6x^2 - 2$

  D $4x^2$

  E $6(x^2 - 1)$

7. Which of the following is equivalent to $3x - (2 + 6x) \leq 28$?

  A $x \leq -10$

  B $-3x \leq 30$

  C $-3x + 2 \leq 32$

  D $-3x - 2 + 6x \leq 28$

  E $6x - (4 + 12x) \leq 14$

Grade 9

8. The product of $3\begin{bmatrix} 4 & -1 \\ 2 & 0 \\ -3 & 5 \end{bmatrix}$ is the same as:

A $\begin{bmatrix} 7 & 2 \\ 5 & 3 \\ 0 & 8 \end{bmatrix}$

B $\begin{bmatrix} 12 & -1 \\ 6 & 0 \\ -9 & 5 \end{bmatrix}$

C $\begin{bmatrix} 12 & -3 \\ 6 & 0 \\ -9 & 15 \end{bmatrix}$

D $\begin{bmatrix} 4 & -3 \\ 2 & 0 \\ -3 & 15 \end{bmatrix}$

E $\begin{bmatrix} 12 & 3 \\ 6 & 0 \\ 9 & 15 \end{bmatrix}$

9. A swimming pool charges a $75 membership fee per year, and $1.50 each time you bring a guest. Which equation shows the yearly cost y in terms of the number of guests g?

A $y = 75g + 1.5$

B $y = 1.5g + 75$

C $y = -1.5g + 75$

D $y = 1.5g + 75g$

E $y = 1.5g - 75g$

10. The table shows t, the charges in cents, for a long-distance phone call that lasts m minutes.

t	1	2	3	4
m	17	26	35	44

Which equation describes this relationship?

A $m = 17t$

B $m = 9t + 8$

C $m = 17t + 9$

D $m = 9t$

E $m = 8t + 9$

11. Which expression is equivalent to $-3(2x - y^2)$?

A $-3xy^2$

B $9xy^2$

C $6(xy)^2$

D $-5x - 6y$

E $-6x + 3y^2$

12. The equation $y = 3x + 7$ is equivalent to which of the following?

A $-3x + y = 7$

B $\dfrac{-3x + y}{3} = -7$

C $3x + y = -7$

D $y = \dfrac{-3x + 7}{-1}$

E $x + y = 7 - 3$

13. The expression $\dfrac{-b^2 a^2 s}{-2a^{-3}s^2}$ has the same value as:

A $\dfrac{b^2 a^5}{2s}$

B $\dfrac{b^2 s^3}{2a^{-5}}$

C $\dfrac{2b^2 a}{s}$

D $19,823$

E $\dfrac{2a^{-3}s^2}{b^2 a^2 s}$

Test Preparation Practice
Algebra

> **12.5.4.a** Solve linear, rational, or quadratic equations or inequalities.

Solve each problem, and circle the letter of the best answer.

1. Which real number do you add to both sides of the equation $3x - 8 = 12x + 1$ so that the equivalent equation is $3x = 12x + 9$?

   A -8

   B -1

   C 1

   D 8

   E 12

2. A rectangular picture frame has a perimeter of 28.5 cm. The width of the frame is $\frac{1}{2}$ of the length. Which equation can be used to determine the length of the picture frame?

   A $4\ell = 28.5$

   B $2.5\ell = 28.5$

   C $2\ell + \frac{1}{2}\ell = 28.5$

   D $2\ell + 2\left(\frac{1}{2}\ell\right) = 28.5$

   E $2\ell + 1\ell = 25.5$

3. Solve: $4y - 3(y + 8) = 12$

   A $y = 5$

   B $y = 8$

   C $y = 12$

   D $y = 36$

   E $y = 24$

4. Which operation is used on both sides of the equation to solve it?

   $$\frac{3}{4}x = 24$$

   A Add $\frac{3}{4}$.

   B Subtract $\frac{3}{4}$.

   C Multiply by $\frac{3}{4}$.

   D Multiply by $\frac{4}{3}$.

   E Divide by 3.

5. Solve: $x + \frac{1}{2} < \frac{x}{4} + 2$

   A $x < \frac{1}{2}$

   B $x < 2$

   C $x > 2$

   D $x > 3$

   E $x < 3$

6. What value must be added to both sides of the equation $x^2 - 12x = 15$ in order to solve by completing the square?

   A 15

   B 36

   C 51

   D 144

   E 195

Grade 9

7. Solve for c: $bc + 2ac = 12$

   A $c = 12 - b - 2a$

   B $c = 12 + b + 2a$

   C $c = \dfrac{12}{b + 2a}$

   D $c = \dfrac{12}{2a}$

   E $c = \dfrac{6}{2a}$

8. Which method can NOT be used to solve the quadratic equation $x^2 + 8x + 7 = 0$?

   A Factoring

   B Quadratic formula

   C Taking the square root

   D Completing the square

   E All methods can be used

9. Solve: $2x^2 - 6x + 7 = -4x + 2x^2$

   A $x = -3\dfrac{1}{2}$

   B $x = 3\dfrac{1}{2}$

   C $x = \dfrac{7}{10}$

   D $x = -\dfrac{7}{10}$ or $x = 3\dfrac{1}{2}$

   E $x = -\dfrac{3}{2}$ or $x = \dfrac{10}{7}$

10. The length of an indoor youth soccer field is 2.75 times the width. The field will use 46,475 square feet. Which equation can be used to find the width x of the field?

   A $3.75x^2 = 46{,}475$

   B $2.75x^2 = 46{,}475$

   C $x = \pm\sqrt{2.75 \cdot 46{,}475}$

   D $\dfrac{x^2}{275} = 46{,}475$

   E $x^2 = 46{,}475 + 3.75$

11. Solve: $15x + 8 = 12x - 7$

   A $x = -5$

   B $x = -\dfrac{1}{3}$

   C $x = \dfrac{5}{9}$

   D $x = \dfrac{3}{4}$

   E $x = 2$

12. Gillian bought a dozen eggs for $3.25 and 6 mangos for another price. Her total bill before sales tax was $8.05. What was the cost of each mango?

   A $0.50

   B $0.60

   C $0.75

   D $0.80

   E $0.90

Grade 9

Test Preparation Practice
Algebra

> **12.5.4.c** Analyze situations or solve problems using linear or quadratic equations and inequalities symbolically or graphically.

Solve each problem, and circle the letter of the best answer.

1. Describe the solution of $3x \le 21$.

   A All real numbers less than 7

   B All real numbers less than or equal to 7

   C All real numbers greater than or equal to 7

   D All real numbers less than 18

   E All real numbers greater than 7

2. Which graph represents the solution of $x + 6 > 8$?

   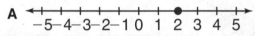

   A
   ```
   -5-4-3-2-1 0 1 2 3 4 5
   ```

   B
   ```
   -5-4-3-2-1 0 1 2 3 4 5
   ```

   C
   ```
   -5-4-3-2-1 0 1 2 3 4 5
   ```

   D
   ```
   -5-4-3-2-1 0 1 2 3 4 5
   ```

   E
   ```
   -5-4-3-2-1 0 1 2 3 4 5
   ```

3. Which of the following is the solution to the equation $4x + 8 = 24$?

   A 2

   B 4

   C 8

   D 16

   E 48

4. The number line below represents the solution to which equation?

   ```
   -5-4-3-2-1 0 1 2 3 4 5
   ```

   A $|w| = -1$

   B $|w + 3| = 2$

   C $|w| = 5$

   D $|w + 7| = 2$

   E $|w - 1| = -5$

5. Use the diagram of the rectangle to find the value of x.

   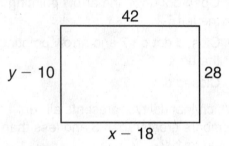

   A 21

   B 24

   C 38

   D 42

   E 60

6. Kelly's class is collecting box tops. Kelly has collected 75 and plans to collect an additional 4 box tops each week. The equation $y = 75 + 4x$ can be used to represent this situation. How many weeks will it take her to collect 119 box tops?

   A 8 weeks

   B 11 weeks

   C 15 weeks

   D 44 weeks

   E 48 weeks

Grade 9

7. Which graph represents the solution of
$-6 < x - 3 \le -1$?

A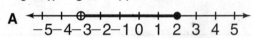
-5-4-3-2-1 0 1 2 3 4 5

B
-5-4-3-2-1 0 1 2 3 4 5

C
-5-4-3-2-1 0 1 2 3 4 5

D
-5-4-3-2-1 0 1 2 3 4 5

E
-5-4-3-2-1 0 1 2 3 4 5

8. Describe the graph of the solution of
$6x = 42$.

A Open dot on 7 and arrow pointing to the right

B Closed dot on 7 and arrow pointing to the left

C Open dot on 7 and arrow pointing to the left

D Closed dot on 7 and arrow pointing to the right

E Closed dot on 7

9. Which inequality represents all real numbers greater than 5 and less than or equal to 12?

A $5 < x \le 12$

B $5 \le x \le 12$

C $5 < x < 12$

D $5 > x$ or $x \le 12$

E $5 < x$ or $x \ge 12$

10. Your soccer team averages between 1 and 5 goals per game. Which absolute-value inequality describes the average number x of goals per game?

A $|x - 2| \le 5$

B $|x - 3| \le 5$

C $|x - 3| \le 2$

D $|x + 2| \ge 5$

E $|x - 2| \le 5$

11. Which graph represents the solution of the inequality?

$4x + 1 < 17$ or $3x - 11 \ge 10$

A
2 3 4 5 6 7 8 9

B
2 3 4 5 6 7 8 9

C
2 3 4 5 6 7 8 9

D
2 3 4 5 6 7 8 9

E
2 3 4 5 6 7 8 9

12. A bag of rice lists a possible cooking time of 22 to 25 minutes. Which inequality represents the possible cooking times?

A $22 < x \le 25$ **D** $3 \le x$

B $22 \le x \le 25$ **E** $22 > x > 25$

C $22 > x \le 25$

13. Portia is in charge of buying sandwiches for the yearbook's club meeting after school. She has $34, and sandwiches cost $4.25 each. How many sandwiches can she buy?

A 4 sandwiches   **D** 11 sandwiches

B 6 sandwiches   **E** 12

C 8 sandwiches

14. The Daniels family is taking a vacation to Niagara Falls. They want to save at least $550 before their trip for souvenirs. Which number line shows the amount of money that the Daniels family would like to save?

A
550

B
550

C
550

D
550

E
550

Grade 9

Test Preparation Practice
Algebra

12.5.4.d Recognize the relationship between the solution of a system of linear equations and its graph.

Solve each problem, and circle the letter of the best answer.

1. Which of the following ordered pairs indicates the solution of the system of equations in the graph below?

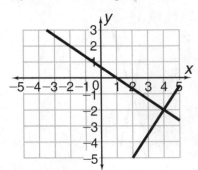

A $(-4, 2)$ **D** $(-2, 4)$

B $(2, 4)$ **E** $(4, -2)$

C $(4, 2)$

2. The graph below is a graph of which system of equations?

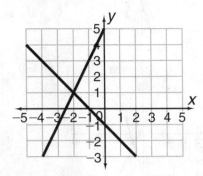

A $2x = y + 5$
   $x + y = 1$

B $2x = y - 5$
   $x + y = -1$

C $x = y + 5$
   $y = 1 - x$

D $2x - y = -5$
   $x + y = -1$

E $y - 3x = 5$
   $x - 1 = y$

3. Which system of inequalities is represented by the graph?

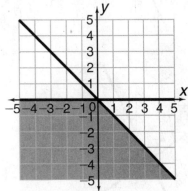

A $x - y < 0$
   $y \leq 0$

B $x + y < 0$
   $y \leq 0$

C $x - y < 0$
   $y \geq 0$

D $x + y > 0$
   $y \leq 0$

E $x - y > 0$
   $y \geq 0$

4. What do the lines $2x - y = -5$ and $x + y = 5$ have in common?

A the y-intercept

B the x-intercept

C the y-value when $x = 4$

D the slopes

E the axis of symmetry

Grade 9

5. Which graph represents the solution to the system:
$$-2x = y$$
$$2x - 3y = -16$$

A

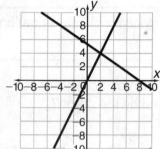

B

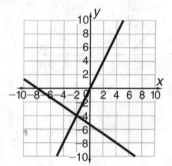

C

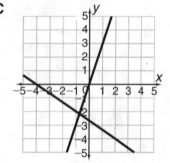

D

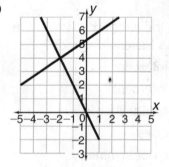

E

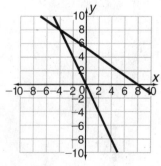

6. Which of the following is the graph of
$$y \geq -4x + 1$$
$$y \leq 3x - 5$$?

A

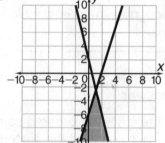

B

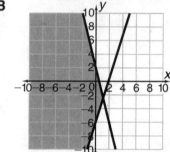

C

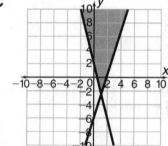

D

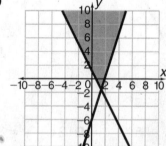

E

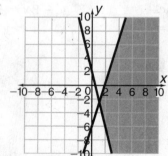

Grade 9

Test Preparation Practice
Algebra

12.5.4.f Given a familiar formula, solve for one of the variables.

Solve each problem, and circle the letter of the best answer.

1. Your cell phone bill is based on a flat monthly fee, and the number of minutes used. In the equation $C = 0.07m + 29.99$, what does the variable m represent?

   A The number of months billed

   B The total amount of the bill

   C The number of minutes used

   D The phone number

   E The cost for using the minutes

2. The band is trying to raise money so that they can take a field trip to the Rock and Roll Hall of Fame. They decide to sell sweatshirts. The equation for the amount of money (a) that they will make for selling t sweatshirts is $a = 22t - 350$. In order to make at least $2100, how many sweatshirts do they need to sell?

   A 79

   B 80

   C 100

   D 111

   E 112

3. The bus company in town sells a bus pass each year for $8.50, and then charges $0.25 a ride. Using the function $f(b) = 0.25b + 8.5$, how much will it cost someone to ride the bus 38 times?

   A $9.50

   B $18.00

   C $20.00

   D $9.50

   E $118.00

4. The volume V of a cylinder is $V = \pi r^2 h$, where r equals the radius and h equals the height. Solve the equation for h.

   A $\dfrac{V\pi}{r^2} = h$

   B $\sqrt{V\pi r} = h$

   C $(V\pi r)^2 = h$

   D $\dfrac{V}{\pi r^2} = h$

   E $V\pi r^2 = h$

5. On a certain standardized test your score is a function of how many questions you answer correctly, using the equation $f(x) = 9x + 218$, where x is the number of questions answered correctly. If the maximum score on the test is a 650, how many questions are there on the test?

   A 96 questions

   B 72 questions

   C 48 questions

   D 24 questions

   E 12 questions

6. Given that $C = \dfrac{5}{9}(F - 32)$, what equation represents F?

   A $\dfrac{9}{5}C - 32 = F$

   B $1.8C + 32 = F$

   C $32C - \dfrac{5}{9} = F$

   D $\dfrac{5}{9}C + 32 = F$

   E $1.8C \times 32 = F$

Grade 9

Which describes this relationship?

A $m = 17t$

B $m = 9t + 8$

C $m = 17t + 9$

D $m = 9t$

E $m = 8t + 9$

7. The Aviation Express company offers a plane ride over your house for $35 plus $2.50 per minute. Using the function $f(r) = 2.50m + 35$, how much will it cost someone to fly over their house if it is a 20 minute flight?

A $35.00

B $37.50

C $85.00

D $87.50

E $122.50

8. The distance traveled by a falling object is given by the formula $d = 4.9t^2$, where d is in meters and t is in seconds. If a diver dives off a 10-m platform, how far will she have fallen in $1\frac{1}{4}$ seconds?

A 4.9 m

B 6.4 m

C 7.7 m

D 9.2 m

E 10 m

9. The function $y = 31,000(0.98^x)$ models the population of a city x years from now. What is the most accurate prediction for the population of a city 5 years from now? Round to the nearest whole number

A 26,000

B 27,231

C 28,022

D 32,167

E 151,900

10.

Rate Plan	Number of Included Minutes	Cost Per Month	Each Additional Minute
A	250	$12.95	$0.35
B	500	$24.95	$0.32
C	750	$32.95	$0.30

Rebecca is looking at Rate plan B for her cell phone. She has determined that the monthly cost can be modeled by the function $c = \$24.95 + \$0.32m$, where c is the total cost of her monthly bill and m is the number of minutes over the included amount. How much would it cost Rebecca if she used her phone 612 minutes?

A $35.84

B $60.79

C $74.63

D $195.84

E $220.79

11. If $V = s^3$, what is s?

A $s = 3V$

B $s = \dfrac{V}{3}$

C $s = \sqrt[3]{V}$

D $s = V^3$

E $s = \sqrt{V}$

Test Preparation Practice
Algebra

> **12.5.4.g** Solve or interpret systems of equations or inequalities.

Solve each problem, and circle the letter of the best answer.

1. To solve the following system of equations by elimination, which operation would you perform first?

   $2.45x + 0.65y = 9.95$
   $6.25x + 0.65y = 21.35$

   A Addition

   B Subtraction

   C Multiplication

   D Division

   E Finding a square root

2. At what point (x, y) do the two lines $x - 2y = 5$ and $3x - 5y = 8$ intersect?

   A $(-9, 7)$

   B $\left(2, -\dfrac{3}{2}\right)$

   C $\left(-2, -\dfrac{7}{2}\right)$

   D $(23, 7)$

   E $(-9, -7)$

3. Choose the statement that is true for a system of two linear equations.

   A A system can only be solved by graphing the equations.

   B There are two solutions when the graphs of the equations are perpendicular lines.

   C There are infinitely many solutions when the graphs of the equations have the same slope and intercepts.

   D There is exactly one solution when the graphs of the equations are one line.

   E When solving a system by substitution the first step is to add the equations.

4. A farmer raises apples and peaches on 215 acres of land. If he wants to plant 31 more acres in apples than peaches, how many acres of each should he plant? Which system of equations will help determine how many acres of each he should plant?

   A $a + p = 215$
   $a + p = 31$

   B $a - p = 215$
   $a + p = 31$

   C $a + p = 215$
   $a + 2p = 31$

   D $a + p = 215$
   $a - p = 31$

   E $a - p = 215$
   $a - p = 31$

5. What is the value of x in the system of equations?

   $5x = 4y - 7$
   $8y = 6x + 2$

   A -2

   B -3

   C 1

   D 2

   E 3

6. The perimeter of a rectangle is 54 in. The length is 3 in. more than the width. What is the length of the rectangle?

   A 13 in.

   B 15 in.

   C 25.5 in.

   D 28.5 in.

   E 31 in.

Grade 9

7. Rayann bought 4 roses and 3 daisies for $21.53. Jason bought 3 roses and 2 daisies and paid $15.45. What is the cost for one rose and one daisy?

A $2.79

B $2.84

C $3.29

D $5.28

E $6.08

8. Megan has $6000 to invest. She earns $350 in interest by putting part of it in her savings account that pays 3% interest and the remainder in stocks that yield 8%. Which system of equations can be used to determine how much she invests at each rate?

A $x + y = 6000$
$0.3x + 0.8y = 350$

B $x + y = 6000$
$3x + 8y = 350$

C $x - y = 6000$
$0.03x + 0.08y = 350$

D $x + y = 6000$
$0.03x + 0.08y = 350$

E $x + y = 350$
$0.03x + 0.08y = 6000$

9. Find the step at which the first error is made in the following work as the system is solved using substitution.

$2x + y = -5$
$2 - 2y = x$

Use the first equation: $2x + y = -5$

A Substitute for x. $2(2 - 2y) + y = -5$

B Distribute. $4 + 4y + y = -5$

C Combine like terms. $5y + 4 = -5$

D Subtract 4 from both sides. $5y = -9$

E Divide by 5 on both sides. $y = \dfrac{-9}{5}$

10. Your chemistry teacher is preparing lab samples. He has 30% and 80% alcohol solutions. He needs 100 mL of a 50% alcohol solution. How many milliliters of the 30% solution should he mix?

A 20 mL

B 40 mL

C 60 mL

D 80 mL

E 100mL

11. Which ordered pair is a solution of the system $\begin{array}{l} 2x + y = -5 \\ 3x + 5y = -4 \end{array}$?

A $(-3, 1)$

B $(3, 1)$

C $(0, -5)$

D $(2, -1)$

E $(-3, -1)$

12. To solve the following system of equations by elimination, which step would you perform first?

$10a + 6b = 8$
$5a + 3b = 2$

A Add the two equations.

B Write the two equations in standard form.

C Subtract the two equations.

D Multiply both equations by -2.

E Multiply the second equation by -2.

Grade 9

Name _____ Date _____ Class _____

Test Preparation Practice
Sample Test A

Solve each problem. Choose the best answer for each question and record your answer on the Student Answer Sheet.

1. Determine the area of the figure.

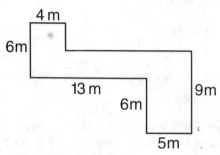

A 60 square meters

B 48 square meters

C 96 square meters

D 111 square meters

E 171 square meters

2. A survey poll was taken one week before an election.

Candidate	Percent of Supporters
Beaver	35%
Lemur	
Other	20%

Write the percentage of those who support Candidate Lemur as a fraction in simplest form.

A $\frac{7}{50}$

B $\frac{1}{5}$

C $\frac{9}{20}$

D $\frac{11}{20}$

E $\frac{2}{3}$

3. What is the volume of a sphere with a radius of 4.5 feet?

$\left(\text{Hint: } V = \frac{4\pi r^3}{3}\right)$

A 56.55 cubic feet

B 254.47 cubic feet

C 381.70 cubic feet

D 563.23 cubic feet

E 1145.45 cubic feet

4. If the speed of light is 3.00×10^8 meters per second, how far would a beam of light travel in 7500 seconds?

A 2.00×10^8 m   D 2.25×10^{12} m

B 2.25×10^8 m   E 2.25×10^{13} m

C 2.25×10^{11} m

5. Maria is a plumber. The amount she charges is based on the number of hours that she works. The total T, in dollars, that she charges is given as a function of the number of hours, h, that she works: $T = 75h + 60$. Which sentence best describes the amount she charges?

A She charges $60 per hour.

B She charges $60 for the first hour and $75 for each additional hour.

C She charges $60 per hour plus a house call fee of $60.

D She charges $60 for the first three hours and $75 for each additional hour.

E She charges $60 per hour plus a house call fee of $75.

Grade 9

6. Using the graph, find the difference between greatest amount of money raised and the least amount of money raised.

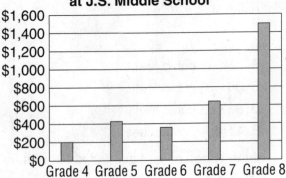

Amount of Fundraising at J.S. Middle School

A. $100

B $1000

C $1200

D $1300

E $1400

7. Joe's commission rate is 8.5%. Last week, his sales were $12,500. What was his commission for the week?

A $1000

B $1062.50

C $1500.65

D $10,625

E $15,000

8. What is the equation of the line that is perpendicular to $y = -2x + 7$ and passes through the point $(-9, -8)$?

A $y = \frac{1}{2}x - \frac{5}{2}$

B $y = 2x - 5$

C $y = \frac{1}{2}x - \frac{7}{2}$

D $y = -\frac{1}{2}x - \frac{7}{2}$

E $y = -2x - \frac{7}{2}$

9. Estimate the length of the bug, in inches.

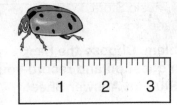

A $\frac{1}{2}$ inch

B 1 inch

C 1.5 inches

D 1.75 inches

E 2 inches

10. What are the coordinates of point A, when the arrow is translated 3 units to the right and 2 units up?

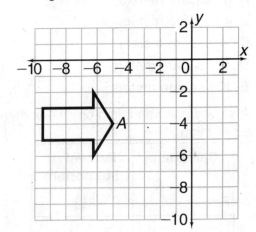

A $(-2, 0)$

B $(-2, -2)$

C $(-5, 2)$

D $(-2, -5)$

E $(0, -5)$

11. Which value is equivalent to $\dfrac{4^{-25}}{4^{-12}}$?

A 4^{-37}

B 4^{-13}

C 4^{-12}

D 4^{12}

E 4^{13}

Grade 9

12. Chuanxi bought the following items at Michael's Music Store. What is the total purchased, rounded to the nearest whole dollar?

Item	Cost
CD	$13.50
DVD	$18.99
batteries	$6.79
telephone	$12.99
poster	$9.99
blank tapes	$5.49

A. $65.00   D $68.00

B $66.00   E $69.00

C $67.00

13. The Jiminez family bought a new washer and dryer on an installment plan. They made 24 monthly payments of $85 and a down payment of $300. How much money would they have saved if they had paid $1990 in cash?

A $250

B $350

C $400

D $1500

E $1750

14. Mykaela worked the following hours during one week. What is the total number of hours that she worked?

Day	Hours
1	12 hours, 30 minutes
2	5 hours, 15 minutes
3	9.75 hours
4	35 minutes
5	$8\frac{1}{3}$ hours

A 25.26 hr   D 40.23 hr

B 35.85 hr   E 69.00 hr

C 36.42 hr

15. What is the ratio of the area of the envelope to the volume of the box in lowest terms?

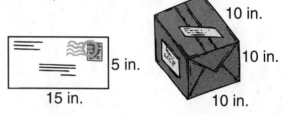

5 in.

15 in.

10 in.

10 in.

10 in.

A 1:40

B 3:40

C 7:40

D 75:100

E 75:1,000

16. There are 15 grams of carbohydrates in a $\frac{3}{4}$-ounce serving of pretzels. How many grams of carbohydrates are in 1 pound of pretzels?

A 100

B 220

C 320

D 420

E 440

17. Mr. Reed is making a miniature bed for a dollhouse using a scale of 0.75 inch to one foot. How long is the original bed if the model is 5.625 inches long?

A 6 feet

B 6.25 feet

C 6.5 feet

D 7 feet

E 7.5 feet

18. A hotel ballroom has 60 red chairs and 84 blue chairs. A wedding planner must put the same number of red and blue chairs at each table. What is the greatest number of tables that can be arranged in this fashion?

A 4   D 12

B 6   E 14

C 10

Grade 9

19. The perimeter of the Jacobsens' yard is 750 feet. What is the length *s*?

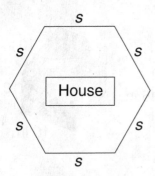

A 100 feet   **D** 125 feet

B 110 feet   **E** 130 feet

C 120 feet

20. The ratio of the corresponding sides of 2 similar triangles is 2:6. The sides of the larger triangle are 12 yards, 15 yards, and 18 yards. What is the perimeter of the smaller triangle?

A 7.5 yd   **D** 25 yd

B 12 yd   **E** 45 yd

C 15 yd

21. What is the length of the diagonal of a square whose sides measure 30 inches?

A 30 inches

B 42.43 inches

C 48.32 inches

D 50 inches

E 52.94 inches

22. Big State University has 23,962 undergraduates, and Small State College has 2136 students. What is the best estimate of the total number of students in the two colleges?

A 22,000   **D** 25,000

B 23,000   **E** 26,000

C 24,000

23. What is the transformation from figure 1 to figure 2?

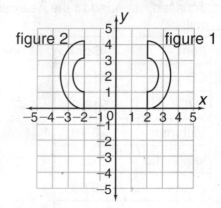

A flip

B slide

C flip, slide

D turn

E turn, slide

24. Jaime buys a tarp to cover the deck of his house. The tarp measures 15 feet by 15 feet. The deck measures 20 feet by 23 feet. How many square feet of the deck is not covered by the tarp?

A 225 ft^2

B 235 ft^2

C 420 ft^2

D 460 ft^2

E 485 ft^2

25. Find the value of *x* that makes the two right triangles similar.

 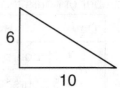

A. 4.2

B. 4.3

C. 4.4

D. 4.5

E. 4.6

Grade 9

Name _____   Date _____   Class _____

Test Preparation Practice
Sample Test B

Solve each problem. Choose the best answer for each question and record your answer on the Student Answer Sheet.

1. What expression comes next in this pattern?

   $4x, 16x^3, 64x^5, 256x^7, \ldots$

   A $1024x^7$

   B $512x^7$

   C $1024x^9$

   D $512x^9$

   E $2048x^7$

2. Kraiga's class is collecting aluminum cans. Kraiga has collected 55 cans and she plans to collect 15 additional cans per week. The equation $y = 55 + 15x$ models this situation. How many weeks will it take her to collect 355 cans?

   A 10

   B 15

   C 20

   D 25

   E 30

Use the data to answer Questions 3 and 4.

The number of community service hours per month for the high school Key Club are given in the following list:

12, 47, 21, 28, 40, 34, 16, 5, 23, 21.

3. What are the first quartile, median, and third quartile of the data?

   A 16, 20, 35

   B 12, 16, 20

   C 16, 22, 34

   D 5, 34, 40

   E 16, 20, 40

4. Find the mean, mode and range, respectively.

   A 24.7, 21, 42

   B 20.5, 21, 34

   C 24.7, no mode, 42

   D 24.7, 21, 55

   E 22.7, 22, 25

5. Which system of inequalities is represented by the graph?

   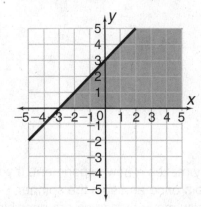

   A $x - y \le 3$
      $y > 3$

   B $x - y \le 3$
      $y = 0$

   C $x - y \ge -3$
      $y > 0$

   D $x - y = 3$
      $y > 0$

   E $x - y > -3$
      $y < 0$

6. The teaching assistant for Chemistry 101 is preparing a solution mixture for demonstration. She has 20% and 60% alcohol solutions. She needs 50 mL of a 50% alcohol solution. How many milliliters of the 60% solution does she need to mix with the 20% solution?

   A 12.5 mL   **D** 37.5 mL

   B 25.5 mL   **E** 50 mL

   C 33.5 mL

Grade 9

Use the spinner to answer Questions 7 and 8.

7. A boy spins the spinner for a game. What is the probability that it lands on a shaded area and a consonant?

A $\frac{1}{9}$

B $\frac{2}{9}$

C $\frac{3}{9}$

D $\frac{5}{9}$

E $\frac{7}{9}$

8. The boy makes a new experiment. If you flip a coin once and spin the spinner once, how many outcomes are possible?

A 10

B 11

C 13

D 18

E 30

9. Christoph has 9 more baseball trading cards than Miguel. The number of cards that they have combined is 45. Which equation represents how many cards, c, Miguel has?

A $2c + 9 = 45$

B $c + 9 = 45$

C $c = 45$

D $3c + 9 = 45$

E $2c + 9 = 45 - c$

10. What is the function graphed below?

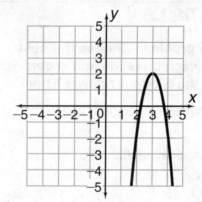

A $f(x) = -4x^2$ shifted 2 units up and 3 units right.

B $f(x) + 2 = -4x^2 - 3$

C $f(x) = -4x^2$ shifted 3 units up and 3 units left.

D $f(x) = -4x^2 + 3$

E $f(x) = -4x(x + 2)^2 - 3$

Use the information to answer Questions 11 – 14.

A jar has 5 red, 7 blue, and 3 yellow buttons. The red buttons are numbered 1–5. The blue buttons are numbered 2–8, and the yellow buttons are numbered 21–23. One button is drawn at random.

11. What is the probability of NOT drawing a yellow button?

A $\frac{1}{5}$ **D** $\frac{4}{5}$

B $\frac{2}{5}$ **E** $\frac{8}{9}$

C $\frac{3}{7}$

12. What is the probability of drawing a blue button or a yellow button?

A $\frac{1}{3}$ **D** $\frac{4}{5}$

B $\frac{2}{3}$ **E** $\frac{14}{15}$

C $\frac{2}{5}$

Grade 9

13. What is the probability of drawing an odd number?

A $\frac{1}{4}$ **D** $\frac{8}{15}$

B $\frac{1}{3}$ **E** $\frac{3}{4}$

C $\frac{7}{15}$

14. What is the probability of drawing an even number?

A $\frac{1}{3}$ **D** $\frac{3}{4}$

B $\frac{7}{15}$ **E** $\frac{8}{15}$

C $\frac{3}{5}$

15. The Open Air company charges a flat fee of $75, plus $15 per hour, for a hot air balloon ride. Josephine can spend $150 on a ride. How long will her trip be?

A 2 hours

B 3 hours

C 5 hours

D 6 hours

E 8 hours

16. Solve for c and x.

$$\begin{bmatrix} 9+t & 3c \\ d-31 & 5-x \end{bmatrix} = \begin{bmatrix} 10 & 12 \\ -7 & -15 \end{bmatrix}$$

A $c = 2, x = 20$

B $c = 4, x = 10$

C $c = 4, x = 20$

D $c = -4, x = 10$

E $c = -4, x = 0$

17. What is the pattern?

$16, -4, 1, -\frac{1}{4}, \frac{1}{16}, \ldots$

A Divide by 4.

B Divide by -4.

C Multiply by 4.

D Multiply by 16.

E. Multiply by -4.

Use the circle graph to answer Questions 18 and 19.

McKay Family Investment Portfolio

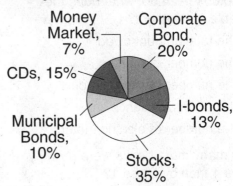

18. Which type of investment is the McKay family's largest?

A corporate bonds

B I-bonds

C stocks

D CDs

E municipal bonds

19. The family investment portfolio is currently worth $625,000. How much are the corporate bond holdings worth?

A $62,500   **D** $145,000

B $125,000   **E** $340,000

C $130,000

20. The ordered pairs shown form a quadratic pattern.

x	y
0	2
1	5
2	14
3	29
4	50
5	??

What is the missing value of y?

A 49   **D** 77

B 63   **E** 80

C 65

Grade 9

21. Your electric bill is based on a flat monthly fee and the number of kilowatt hours used. In the equation $c = 0.05k + 24.50$, what does the variable k represent?

A The total amount owed

B The number of months

C The number of kilowatt hours

D The account balance

E The number of electric meters

22. How many more books were read by Class 4 than by Class 1?

Class	Number of Books Read (Legend 📖 = 20)
1	📖 📖
2	📖 📖 📖
3	📖 📖 📖 📖 📖
4	📖 📖 📖

A 10

B 15

C 20

D 30

E 40

Use the graph to answer Questions 23 and 24.

The graph gives the value of Kristine's color copier since she purchased the copier.

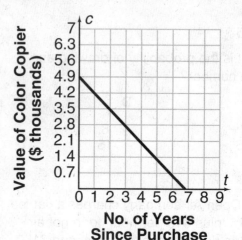

No. of Years
Since Purchase

23. What is the equation of the line?

A $c = -7t + 490$

B $c = 7t - 4.9$

C $c = -0.7t + 4.9$

D $c = 0.7t + 49$

E $c = -7t + 49$

24. How much was the copier worth 4 years after Kristine purchased the copier?

A $200

B $1800

C $2100

D $2400

E $3000

25. Evaluate $x^5 - 23x^2 + 45x - 34$ for $x = 2$.

A -10

B -6

C -4

D 0

E 5

Name _____ Date _____ Class _____

Sample Tests
Answer Sheet

1 Ⓐ Ⓑ Ⓒ Ⓓ Ⓔ   14 Ⓐ Ⓑ Ⓒ Ⓓ Ⓔ

2 Ⓐ Ⓑ Ⓒ Ⓓ Ⓔ   15 Ⓐ Ⓑ Ⓒ Ⓓ Ⓔ

3 Ⓐ Ⓑ Ⓒ Ⓓ Ⓔ   16 Ⓐ Ⓑ Ⓒ Ⓓ Ⓔ

4 Ⓐ Ⓑ Ⓒ Ⓓ Ⓔ   17 Ⓐ Ⓑ Ⓒ Ⓓ Ⓔ

5 Ⓐ Ⓑ Ⓒ Ⓓ Ⓔ   18 Ⓐ Ⓑ Ⓒ Ⓓ Ⓔ

6 Ⓐ Ⓑ Ⓒ Ⓓ Ⓔ   19 Ⓐ Ⓑ Ⓒ Ⓓ Ⓔ

7 Ⓐ Ⓑ Ⓒ Ⓓ Ⓔ   20 Ⓐ Ⓑ Ⓒ Ⓓ Ⓔ

8 Ⓐ Ⓑ Ⓒ Ⓓ Ⓔ   21 Ⓐ Ⓑ Ⓒ Ⓓ Ⓔ

9 Ⓐ Ⓑ Ⓒ Ⓓ Ⓔ   22 Ⓐ Ⓑ Ⓒ Ⓓ Ⓔ

10 Ⓐ Ⓑ Ⓒ Ⓓ Ⓔ   23 Ⓐ Ⓑ Ⓒ Ⓓ Ⓔ

11 Ⓐ Ⓑ Ⓒ Ⓓ Ⓔ   24 Ⓐ Ⓑ Ⓒ Ⓓ Ⓔ

12 Ⓐ Ⓑ Ⓒ Ⓓ Ⓔ   25 Ⓐ Ⓑ Ⓒ Ⓓ Ⓔ

13 Ⓐ Ⓑ Ⓒ Ⓓ Ⓔ

Grade 9